AI Made Simple. Results Made Real.

An Executive's Guide to Partnering with the Future

by Kathleen Perley

Kingbird
PRESS

Published by Kingbird Press
www.kingbirdpress.com

ISBN:
978-1-965629-02-4 (Hardback)
978-1-965629-01-7 (Paperback)
978-1-965629-03-1 (eBook)
978-1-965629-04-8 (Audiobook)

Cover design by Jeff Heaton
Layout by Becky Noll Reinert

AI-assisted tools were used in the creation of this book.

In memory of my mom, Eileen Campbell, whose unwavering belief transformed every "you can't" into a powerful "just watch me." Your ability to see beauty in my differences—curiosity, creativity, and a talent for patterns—is the heart of this book. May this work inspire others to embrace their unique strengths, especially in this age of AI, where different thinkers now thrive.

CONTENTS

CONTENTS

Foreword by Marie Meyer

My journey into artificial intelligence didn't begin from the vantage point of a technologist. Instead, I entered this field through business, believing technology works best when guided by a clear strategy and thoughtful leadership.

Having spent decades in finance—from my early days at Compaq in Houston to my current role as CFO at Hewlett Packard Enterprise—I learned early that my non-technical perspective provided useful insights. Working with a startup focused on Robotic Process Automation (RPA) further opened my eyes to AI's possibilities. It reinforced an important idea: innovation happens when business experience meets new technology.

Alongside my colleague and friend Kathleen Perley, I've had the chance to co-teach a course on AI at the Jones Graduate School of Business at Rice University. This experience shaped how we think about preparing today's business leaders with practical AI knowledge and reinforced the urgent need for executives to get a "minor" in artificial intelligence to remain relevant and impactful. Our collaboration led to an exciting AI finance project between Hewlett Packard Enterprise and Kathleen's company, DemystifAI. Her skill at simplifying complex ideas into clear, practical business strategies is exactly what's needed amid the rapid pace of change in AI.

Kathleen's book, "AI Made Simple. Results Made Real: An Executive's Guide to Partnering with the Future," helps leaders navigate these changes. It explains artificial intelligence in simple terms, helping leaders incorporate AI into their organizations. Executives don't need deep technical skills; they need enough understanding to guide their teams, make informed decisions, and encourage a culture open to innovation.

The book also covers important topics like ethics, legal concerns, talent management, and staying competitive. Through real examples and practical frameworks, it highlights AI's transformative potential across business functions—from enhancing customer interactions at T-Mobile to revolutionizing customer service at the Swedish fintech company Klarna, where an AI assistant efficiently managed millions of customer interactions in over 35 languages, significantly reduced query resolution times, and was projected to deliver a $40 million profit improvement in 2024.

I invite you, as business leaders, to embrace the visionary role demanded by this AI-driven era. Equipped with the roadmap provided here, you'll be prepared to lead your organization toward a future marked by thoughtful innovation and steady growth.

Marie Meyer
Chief Financial Officer of Hewlett Packard Enterprises

INTRODUCTION

There's no denying it: Artificial intelligence is no longer a distant concept. It is fundamentally reshaping industries, redefining competition, and reimagining how businesses operate - TODAY. Companies that fully integrate AI are growing their revenue 50% faster than their competitors *(Accenture, 2023)*, making AI adoption a competitive requirement. While AI's long-term economic impact is projected to reach $13 trillion by 2030, its impact is already clear, with over 50% of organizations actively using AI in at least one business function *(McKinsey, 2023)*. Despite its immense potential, many business leaders struggle to determine how AI fits into their operations, how to apply it effectively, and what it means for their organizations today.

AI in Action

When I speak with businesses or present at conferences, I often find that many attendees have used AI for tasks like drafting emails or for a fun party trick. However, most don't fully grasp the breadth of AI's capabilities, as AI does not just generate text and analyze data. It can see, speak, reason, and adjust to changes in real life (like customer requests or real-world interactions) that it has not been programmed to address.

Before diving into the fundamentals, I want to expand your perspective on what AI can do and how it can be integrated into your organization in ways you may not have considered. To illustrate this, I'll share a few case studies that highlight some of AI's lesser-known but highly impactful applications.

Example 1: AI Avatars in Multiple Languages

Heineken, the globally recognized beer company, embraced AI video technology to enhance its internal training and communications. Faced with the challenge of providing consistent training content for its 70,000 employees worldwide, Heineken collaborated with an AI video platform to produce engaging, multilingual training videos.

What They Did:

- **AI-Powered Avatars:** Lifelike digital presenters delivered training content with natural speech, facial expressions, and lip-syncing, creating a consistent, engaging experience.

- **Multilingual Scaling:** Text-to-speech and machine translation enabled instant adaptation of videos into multiple languages, eliminating the need for costly, region-specific filming.

- **Streamlined Production:** Automating script adaptation and video creation reduced production time by 90%, slashing costs and accelerating content delivery.

Real-World Impact:

- **Higher Engagement:** AI-generated content kept employees more engaged with training materials.

- **Cost and Time Efficiency:** Significant reductions in production costs and turnaround times enabled rapid updates.

- **Global Consistency:** Training delivered in employees' native languages led to better comprehension and compliance.

The Bigger Picture

Heineken's approach demonstrates that AI isn't merely about operational successes; it's about unlocking new opportunities. From breaking down language barriers to fostering more human-like interactions, AI-powered tools aren't replacing connection; they're enhancing it. Companies that embrace AI this way are broadening what's possible, making communication more inclusive, training more engaging, and global teams more connected than ever before.

Example 2: Computer Vision is Improving Safety

A leading automotive parts supplier deployed Chooch's Vision AI to reduce forklift–pedestrian collisions, which previously accounted for 34% of all warehouse injuries. With fast-paced operations and frequent human-machine interactions, traditional safety measures like warning signs and manual supervision were ineffective in preventing accidents. The company integrated Chooch's AI-powered computer vision system with its existing CCTV cameras and IoT sensors to create a multi-layered proactive safety net. AI continuously monitored warehouse activity, tracking both forklifts and pedestrians to prevent collisions before they happened.

What They Did:

- **Proximity Alerts:** AI monitors forklift speed and pedestrian movement in real time, activating safety warnings through onboard displays and floor LEDs.

- **Automated Access Control:** AI enforces geofenced safety zones by automatically disabling forklifts when they enter pedestrian-only areas.

- **Predictive Risk Analysis:** The system examines movement patterns to identify high-risk zones and propose workflow adjustments to reduce crossing paths.

Real-World Impact:

- **85% Reduction in Near-Miss Incidents:** Intelligent alerts significantly reduced close calls between forklifts and workers.

- **40% Faster Response to Hazards:** Real-time AI dashboards allowed supervisors to intervene immediately when safety risks arose.

- **100% Compliance with Speed Limits:** AI-monitored zones ensured that forklifts followed speed restrictions in high-traffic areas.

The Bigger Picture: AI as an Intelligent Layer for Physical Environments

Chooch's Vision AI system's approach to safety is proactive rather than reactive, because it adapts dynamically to evolving warehouse conditions through advanced computer vision, spatial analytics, and edge computing.

Ultimately, companies that embrace this technology are establishing new standards for industrial safety, driving operational excellence, and protecting their workforce in increasingly complex environments.

Example 3: AI Robotics Optimize Manufacturing

In August 2024, BMW piloted Figure 02 humanoid robots at its Spartanburg, South Carolina, manufacturing plant to enhance efficiency in sheet metal insertion tasks. With human-like dexterity and multi-modal AI perception, these robots addressed labor shortages and improved precision in complex assembly processes. Traditional automation struggled to provide the flexibility required for such tasks, making these robotics a breakthrough solution.

What They Did:

- **AI-Powered Perception and Dexterity:** Six RGB cameras and NVIDIA-powered vision-language models enabled real-time reasoning. The robots' dexterous hands handled 44-pound payloads with millimeter precision.

- **Autonomous Task Execution:** Smart robots identified sheet metal parts using computer vision, applied force-controlled precision to grasp and align components, and inserted them into chassis fixtures at production-line speeds.

- **Factory Floor Mobility:** The robots navigated the workspace at 2.6 mph, avoiding obstacles and adapting movements based on real-time environmental changes.

Real-World Impact:

- **98% Task Accuracy:** AI-driven automation maintained near-perfect precision, even in dynamic environments.

- **40% Faster Cycle Times:** Robots outperformed manual insertion, eliminating bottlenecks and streamlining production.

- **Ergonomic Risk Reduction:** AI automation took over repetitive overhead tasks, decreasing worker fatigue and injury rates.

The Bigger Picture: AI-Powered Robotics as Workforce Multipliers

BMW's pilot program demonstrates how AI-powered robotics are closing the divide between traditional automation and human labor. By integrating advanced robotics, manufacturers can address labor shortages, boost operational productivity, and enhance workplace safety. As automation improves, the manufacturing industry is set for a future where human expertise and intelligent machines collaborate seamlessly.

Who This Book Is For

This book is designed for executives and business leaders eager to grasp AI's true potential without getting bogged down in technical details. By blending strategy with real-world examples, it will provide leaders with the necessary tools to drive AI adoption.

Through detailed case studies, practical examples, concise explanations, and actionable steps, I aim to offer clear, practical insights and a balanced combination of theory and real-life applications. Whether you are at the beginning of your AI journey or seeking to refine your current strategy, this guide will empower you to understand how AI can be integrated into your strategic planning and drive positive disruption within your organization.

To start, it's crucial to reflect on your organization's current relationship with AI. Through my experience working with various companies, I've identified four common approaches to AI adoption:

- **The Curtain Holders:** These organizations intentionally avoid AI, sticking to traditional methods and missing opportunities for advancement.

- **The Audience Watchers:** These companies prefer to wait and watch, learning from the early adopters before deciding to experiment with AI.

- **The Spotlight Seekers:** Eager to innovate, these businesses are quick to experiment, iterate, and adapt their strategies as they learn from successes and setbacks.

- **The Born Performers:** For these companies, AI isn't an add-on; it's embedded at the core of their operations and strategic planning.

While it may be common for organizations to start as Curtain Holders or Audience Watchers, the future will belong to those who actively embrace change, whether as Spotlight Seekers or Born Performers. Successful AI adoption requires a blend of technical know-how, human insight, industry expertise, and strong leadership. Beyond proficiency and automation, AI is reshaping competitive dynamics, customer expectations, and workforce struc-tures, making strategic AI leadership more crucial than ever.

AI is **general** and **foundational**. It will impact all of a company's functions, such as human resources, finance, manufacturing, marketing, research, customer relations, and more. Companies that set AI as their foundation will outperform their non-AI peers in innovation and speed to market. Companies that prefer to wait to implement AI will do so at their own peril.

My Journey to AI

My journey to AI mastery is both personal and professional. Growing up with dyslexia made language a challenge—one that felt overwhelming until I discovered linguistics in college. It's perhaps ironic that a field so rooted in language became my saving grace, but this intersection of words and logic ignited my passion. Linguistics broke language down into logical, mathematical equations, transforming something that had once seemed complex and confus-ing into a series of patterns I could grasp. This breakdown forms the foundation that powers many generative AI technologies, such as large language models (LLMs). My dyslexia also led me to embrace technology; tools like electronic dictionaries and spell-checking features on computers removed barriers to learning and helped me reach my full potential.

Technology is often seen as scary or overcomplicated (because many people make it that way), but it is profound in unlocking new opportunities. When I started my company, DECODE, I wanted to remove the smoke and mirrors often accompanying the digital landscape to help people realize and capture the potential of it. As an entrepreneur who successfully sold a business to private equity in a fast-paced industry, I've combined technical insights with real-world experience. My journey is a testament to how dissecting complex concepts into manageable, actionable steps can empower leaders to drive change.

Just as I learned to navigate and master language through structured analysis, executives today must approach AI with a strategic mindset. Much like language, AI can feel complex and intimidating, but with the proper framework, it becomes a powerful tool for decision-making, innovation, and leadership.

How to Use This Book

This book is structured to guide executives and business leaders through the essential phases of AI adoption.

Section 1: AI's Core Principles: Understanding the Foundation

This section explores the key principles of artificial intelligence, from basic definitions and terminology to ethical dilemmas and security risks. A keen eye on real-world business integration ensures you'll gain the foundational concepts needed to navigate AI's capabilities, limitations, and strategic applications.

Section 2: Your Roadmap to AI Adoption: Process, People and Pilot Projects

To unlock AI's full potential, leaders must embrace practical, actionable strategies for effective implementation. This section explores essential steps, including assessing organizational readiness, establishing robust governance frameworks for ethical AI, and attracting visionary AI leaders and pilot projects. You'll also learn how to initiate pilot projects that are closely aligned with your strategic goals and existing capabilities.

Section 3: The Strategic Horizon: From AI Adoption to Industry Transformation

Optimizing efficiency is a crucial first step in AI adoption, but long-term success requires moving beyond incremental gains to drive revolutionary change. This section examines how to strategically transition from operational enhancements to transformative ecosystems that enable new business models and competitive advantage. You'll discover how specialized and multimodal AI systems, in conjunction with emerging technologies, will reshape industries, address unprecedented challenges, and unlock opportunities that redefine the future of business.

By the end of this book, you'll not only understand AI's role in your business but also have a clear roadmap for integrating it into your organization in a way that drives long-term value.

For a deeper dive into the case studies and concepts discussed throughout this book, please visit:

https://www.demystifai.com/aimadesimple

SECTION ONE

AI's Core Principles:
Understanding the Foundation

CHAPTER 1:
Beginning with the Basics Behind AI

Artificial intelligence is revolutionizing business and society, but its greatest influence comes from those who understand how it works. Many people perceive AI as a complex and intimidating field filled with technical jargon and abstract concepts. This chapter serves as an accessible guide to the foundational terms and ideas that define AI, demystifying the technology and making it more approachable.

In this chapter, you'll learn:

- The definition of artificial intelligence and how it differs from automation.

- Key AI concepts, including machine learning, neural networks, and algorithms.

- How AI understands and processes language.

- The evolution of AI and why it's now a mainstream topic.

- Practical examples of AI in action across industries.

- Why a strong understanding of AI is essential for business and innovation.

With a solid grasp of AI's foundations, you'll be able to navigate its evolution confidently.

Defining AI: More Than Just Smart Machines

If you were to ask leaders in the AI industry to define artificial intelligence, you'd likely receive a variety of responses. One of my favorite definitions comes from Demis Hassabis, the founder of DeepMind at Google, who describes AI as "the science of making machines smart."

At its core, AI is a branch of computer science that enables machines to learn from data, reason through problems, understand natural language, and interpret sensory inputs. To illustrate AI's reasoning capabilities, consider this simple example:

Imagine I say, "The grass is brown." If I ask my three-year-old what that means, she might respond, "The grass is the color brown." While technically true, it misses a more profound implication: The grass is likely dying and needs water. AI's ability to grasp context and apply reasoning demonstrates that it goes beyond simple memorization or rule-following. It's capable of making inferences similar to human thought.

As a self-proclaimed linguistics nerd, I can't help but quote one of my favorite linguists, John Firth, who famously said, "You shall know a word by the company it keeps." This insight is key to understanding large language models (LLMs), which predict the next word in a sequence based on context. These models are not magic; they are grounded in patterns and probabilities. Yet, they offer extraordinary potential when combined with human creativity and intent.

Here's an example showing how ChatGPT's text-to-image tool, DALL-E, understands the underlying meaning of a phrase or idiom. By changing just one word in a sentence, DALL-E generates two dramatically different images without any additional instructions:

"He is hitting the <u>books</u>" "He is hitting the <u>bottle</u>"

In both instances, no additional context was provided, yet AI understood that the intention of the phrase was symbolic rather than literal. These images are emotionally driven, as opposed to direct interpretations that might otherwise depict a person physically striking a book or a bottle with their fist.

This distinction highlights the foundational difference between AI and automation.

AI vs. Automation: Key Differences

Many companies claim to use AI when, in reality, they are simply using automation: a process that follows pre-programmed rules without adapting or learning. While both AI and automation improve productivity, they operate in fundamentally different ways.

Let's use home lighting as an example to bring these concepts to life.

- **Automation:** Setting your smart home lights to turn on and off at the same time every day is automation. It follows a pre-programmed schedule without adjustment.

- **AI-powered systems:** A smart lighting system, like Philips Hue AI, learns your preferences, adjusts brightness based on your daily habits, and even responds to changes in natural light.

This adaptability demonstrates AI's potential to personalize experiences, optimize efficiency, and fuel innovation in ways that go beyond simple automation. Here's an overview of the key differences between automation and AI across various aspects of use.

Aspect	Automation	AI
Creation	Created by humans with specific paths and set instructions.	Builds on basic instructions and rules but can reason to solve complex problems on a large scale.
Adaptability	Operates on fixed rules and doesn't adapt or improve unless humans update the rules.	Adapts methods and strategies in real time based on new data.
Effort	Requires a lot of manual effort.	Reduces manual effort by automating processes and learning over time.
Decision-making	Does not provide strategic recommendations; follows predefined workflows.	Recommends the best strategies and plans to pursue next based on performance.

How AI Transformed Everyday Vocabulary

Many AI terms that were once reserved for research labs are now part of boardroom conversations, startup pitches, and even casual dinner table discussions.

The AI Sphere: From General to Specialized

The diagram visualizes how each specialized area of AI builds upon more general technologies. It shows that as you move inward—from the outer circle representing general AI to the inner circles representing deep learning and generative AI—the technologies become more focused and application-specific.

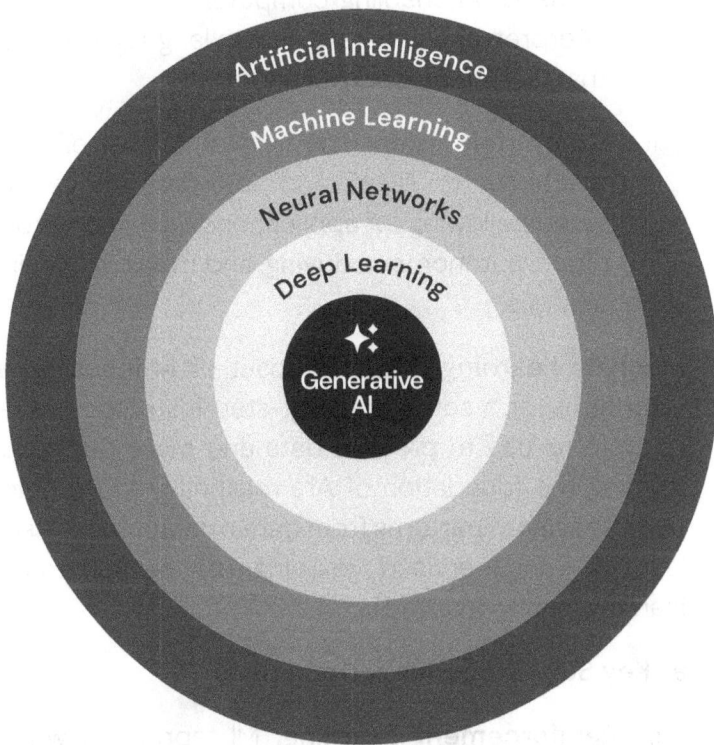

For executives, this layered approach clarifies how foundational concepts like neural networks and machine learning evolve into sophisticated tools like large language models and transformers, which now drive everyday innovations in business and technology.

1. **Artificial Intelligence (AI):** The broadest category encompassing technologies that enable machines to mimic human intelligence, including language understanding, robotics, computer vision, and knowledge representation.

   a. **Key Subcategories & Definitions:**

      i. **Natural Language Processing:** The branch of AI focused on enabling computers to understand, interpret, and generate human language in a way that is both meaningful and useful.

      ii. **Computer Vision:** The field of AI that enables machines to "see" and interpret images or videos, allowing for applications like facial recognition, autonomous driving, and medical imaging analysis.

2. **Machine Learning (ML):** A subset of AI focusing on algorithms, or a set of step-by-step instructions that AI systems use to process data and make decisions, forming the foundation of AI's reasoning and predictions, that learn patterns from data through supervised learning, unsupervised learning, or reinforcement learning (reward-based).

   a. **Key Subcategories & Definitions:**

      i. **Reinforcement Learning:** ML approach where

agents learn optimal behaviors through trial-and-error interactions with an environment, receiving rewards or penalties for their actions.

ii. **Supervised Learning:** ML approach where algorithms learn from labeled data, making predictions based on known examples with correct answers.

iii. **Unsupervised Learning:** ML techniques that find patterns and structure in data without pre-existing labels, often used for clustering, dimensionality reduction, and anomaly detection.

3. **Neural Networks:** A powerful family of algorithms within ML inspired by the human brain's structure. They excel at complex pattern recognition tasks like image processing, speech recognition, and language tasks.

4. **Deep Learning:** A specialized form of neural networks with multiple "deep" layers that can learn increasingly abstract features. Powers include advanced image recognition, natural language processing, and other complex tasks.

5. **Generative AI:** The newest specialized AI field using advanced neural architectures like transformers and Generative Adversarial Networks (GANs) to generate new content—text, images, audio, and code—based on patterns learned from vast datasets.

   a. **Key Subcategories & Definitions:**

      i. **Generative Adversarial Networks:** Framework

where two neural networks (generator and discriminator) compete against each other, with the generator creating increasingly convincing outputs.

ii. **Large Language Models (LLMs):** Massive neural networks trained on vast text corpora that can understand, generate, and manipulate human language with remarkable fluency and versatility.

iii. **Transformer Architecture:** Neural network architecture that uses self-attention mechanisms to process sequential data in parallel, revolutionizing natural language processing.

Why AI's Language Matters

Understanding these terms is essential for engaging with AI both critically and strategically. Whether you lead a company, build a product, or simply seek to understand the future of work, speaking the language of AI enables you to make informed decisions.

Now that we have established the fundamental terminology, we can shift our focus to the larger picture: how AI evolved from a niche technology into a central force shaping industries, economies, and everyday life.

The Conversation Moves from Conference Rooms to Dinner Tables

While the term "artificial intelligence" was coined in 1956 at a conference at Dartmouth College, it has only recently become a pervasive topic of conversation. I vividly remember the day I realized it had become relevant to everyone and AI was no longer just for linguists, engineers, and mathematicians.

It was November 2022, and we were having a typical family dinner. As a native Texan, it's no surprise that most of my family—my mother, father, and sister—work in the oil and gas industry. That evening, my dad casually asked, "Have you heard of this Chat GP... something?" For once, I had the opportunity to be the family expert, a moment I savored. I share this story not as part of a therapy session but to illustrate just how far AI's influence has spread; when my 70-something father is bringing it up at the dinner table, you know it's officially everywhere.

How did we get here? Surprisingly, it wasn't due to a sudden, groundbreaking technological discovery. The real breakthrough happened in 2017 with the release of the Transformer model, a deep learning architecture that fundamentally changed how AI processes language.

Before Transformers, AI struggled with understanding long-form text. Traditional models, like recurrent neural networks and long short-term memory networks, processed words sequentially, meaning they had to read one word at a time in order, like someone listening to a sentence syllable by syllable. This made them slow and prone to forgetting earlier parts of a conversation.

The Transformer model introduced a mechanism called self-attention, which allowed AI to analyze an entire sentence at once rather than word by word. This meant AI could understand context and relationships between words more effectively, even across long passages of text. For example, in the sentence, "The cat, which had been rescued from the storm, sat quietly by the window," a traditional model might struggle to connect "cat" and "sat" because they're far apart. But with self-attention, the Transformer recognizes that "cat" is still the subject, no matter how long the sentence is.

This breakthrough paved the way for today's large language models, including OpenAI's GPT series. The technology itself wasn't new in 2022; what changed was user experience and accessibility. Instead of AI being confined to research labs and specialized applications, tools like ChatGPT placed it directly into the hands of the public. A conversational interface replaced cryptic command lines, making AI as easy to use as texting a friend.

And so, at that dinner table in 2022, my father wasn't marveling at an obscure AI paper from years past. He was reacting to something far more powerful: an AI model that felt intuitive and responsive, making it accessible to anyone for the first time, regardless of technical background.

Driving AI: The Engine, The Controls, and The Key

To explain it further, let's consider a simple analogy: driving a car.

If I had to manipulate the engine directly to get to work or pick up my child from school, I'd never make it. Instead, I use tools like gears, pedals, and a steering wheel to control the car indirectly, guiding it where I need to go. Similarly, picture AI as a powerful engine tucked under the hood of a car. It's the intricate machinery that transforms raw data into predictions and insights, churning with computational horsepower. Yet, much like a driver isn't expected to understand engine mechanics fully, most people who use AI don't need to master the complexities of neural networks. Instead, they rely on the digital equivalent of a steering wheel, pedals, and gears to guide the vehicle without ever peeking inside. This is where the real revolution is taking place: designing interfaces so intuitive that anyone can hop in and drive AI wherever they need to go.

The Key ➡ User Interface (Accessing AI's Power)

Every journey begins with a key. In AI, this represents accessibility—the ability for anyone to unlock and start using powerful AI tools without needing to understand everything under the hood. It's the starting point, the invitation to engage.

The Engine ➡ AI Models (Computational Power)

In our car analogy, the engine provides the raw power. For AI, this engine is the collection of models that perform all the heavy lifting—transforming raw data into predictions, insights, and solutions. These underlying models, like the o3 model from OpenAI, operate much like the engine's intricate mechanics and work tirelessly behind the scenes. Most users don't need to know exactly how these models operate; they simply rely on the performance they deliver.

The Gas Pedal ➡ Inference Speed (Managing Latency)

Just like controlling the gas pedal of a car adjusts acceleration, managing an AI's inference speed influences how quickly or carefully the system responds. Pushing the pedal down—opting for faster reasoning—delivers quicker, real-time answers suitable for straightforward tasks and rapid interactions. Conversely, easing off—choosing slower reasoning—allows the AI to thoughtfully process more intricate problems, improving accuracy and reliability. Balancing this speed and precision ensures the AI's performance aligns effectively with your real-world needs, matching the urgency and complexity of the situation at hand.

The Gear Shift ➡ Model Selection (Task-specific Optimization)

Just as a driver shifts gears to adjust to different driving conditions, the gear shift in AI allows you to change between different models based on the task at hand. Whether you need a model optimized for general writing, advanced logic, image recognition, or another specialized function, shifting gears means selecting the most appropriate model for the job. While we've primarily operated in a manual mode, technology is moving toward an automatic gear shift—one that identifies the best model based on your request without manual intervention.

The Steering Wheel ➡ Prompting (Guiding AI's Output)

Finally, the steering wheel determines the car's direction, just as a carefully crafted prompt guides an AI's response. By thoughtfully adjusting your prompts—like turning the wheel—you set the precise course and focus for the AI, navigating toward the desired outcome. Just as drivers continuously adjust their steering to maintain or correct their path, effective prompting often requires iterative refinements and course corrections. This iterative approach helps you progressively sharpen AI outputs, ensuring they remain aligned with your goals and respond effectively to changing needs or new information along the way.

Putting It All Together

The real revolution in AI isn't about perfecting the engine. It's about creating elegant, intuitive controls that make operating the world's most advanced technology feel as natural as driving a car.

At its core, AI is built on software models. While these models come in various forms, the most foundational are structured around something everyone reading this already understands: the English language. The tools are available; it's just about learning how to drive.

The English-Centric AI Paradox: Progress and Barriers in Multilingual NLP

While today's AI tools are often marketed as "multilingual," their capabilities are far from evenly distributed. Beneath the surface, most large language models (LLMs) remain heavily biased toward English. This imbalance is not just a matter of convenience—it reflects deeper structural and systemic factors that shape the development and deployment of AI technologies.

1. **The Data and Research Advantage:** AI models exhibit remarkable fluency in English primarily due to the vast majority of training data being in English. Over 90% of the text utilized to train foundational models such as GPT-4 and LLaMA 3 originates from English-language sources. This imbalance provides a significant advantage for English and restricts the linguistic versatility of models in other languages.

   Compounding this issue is the fact that AI research is mostly done in English. From academic publications to open-source communities and conferences, English acts as the default lingua franca. This creates a self-reinforcing feedback loop: research advances are tested, benchmarked, and shared in English, which, in turn, influences the design and performance priorities of new models. Languages with fewer digital resources and less representation in research are neglected, both in capabilities and in attention.

2. **English as the "Internal" AI Language:** Recent studies suggest that when processing multilingual inputs, LLMs tend to map information into an abstract "meaning space" that reflects English grammar and syntax.

This internal representation—sometimes referred to as implicit translation—indicates that even when a user inputs text in Spanish, Hindi, or Arabic, the model may effectively translate it into English-like mental structures before generating a response.

This cognitive shortcut introduces subtle yet significant issues. It can distort nuance, cultural idioms, or context-specific expressions that don't translate cleanly into English logic. As a result, AI systems often miss the deeper layers of meaning embedded in other linguistic traditions, leading to outputs that feel stilted, overly literal, or culturally tone-deaf.

3. **Grammar Shapes Machine Understanding:** AI's performance across languages is influenced not only by the quantity of data but also by the inherent complexity and structure of each language. Languages like English, which achieve a balance between structure and flexibility, are relatively easy for models to process. In contrast, languages such as Mandarin—tense-free and context-heavy—or Hungarian—with its rich morphological system—pose significant challenges for tokenization, parsing, and reasoning.

These structural differences go beyond surface difficulty. The way a language encodes time, gender, politeness, or hierarchy can shift the entire interpretive framework required for effective understanding.

On the next page, you'll find a chart taking a comparative look at this.

Aspect	Tense-Free Languages (Mandarin, Indonesian)	Morphologically Rich Languages (Hungarian, Finnish)	Balanced Structure Languages (English)
How time is encoded	Uses context (e.g., "yesterday," "now") instead of verb tenses.	Relies on complex verb endings to indicate time.	Uses verb tenses (past, present, future) combined with contextual clues to convey time clearly yet flexibly.
Grammatical complexity	Minimal verb changes, making it concise and flexible.	Highly structured with multiple suffixes per word.	Moderate grammatical structure, balancing explicit verb tenses with flexible sentence construction.
AI challenges	Struggles with ambiguity—e.g., does "I eat" mean past, present, or future?	Must learn vast word variations—one root word can have hundreds of forms.	Relatively straightforward parsing due to clear tense markers and moderate morphological complexity, aiding computational efficiency.

Understanding these linguistic nuances helps us see beneath AI's polished surface—but admittedly, all this grammar and semantics talk might have you reaching for a cocktail. So, let's use an analogy of crafting the perfect cocktail to truly grasp how AI models are built and function.

The AI Cocktail: Mixing the Perfect Model

Large language models, a cornerstone of modern AI, are built on vast amounts of data. To illustrate how these AI models work, I will share an analogy, so kick back and grab yourself a cocktail. As a proud Texan and margarita lover, I can't help but compare AI to my go-to classic recipe, where tequila, lime, and a pinch of salt come together to create something truly remarkable.

Gathering Ingredients & Bar Tools ➡ Data Curation & Preparation (Foundations of AI Success): Setting up a bar for the evening starts with selecting high-quality ingredients—perhaps a top-shelf tequila, freshly squeezed lime juice, or premium syrups. In AI, this step corresponds to gathering the best possible data. A bartender cannot salvage a drink made with spoiled lime juice, just as an AI model cannot overcome poor-quality or corrupted data, no matter how advanced the algorithm. Think of data preparation as your expert bartender, skillfully choosing and organizing only the best ingredients to ensure each sip—or data point—is perfect.

In real-world AI systems, this might involve deduplicating records, handling missing values, and standardizing formats before training begins—foundational steps that directly impact the quality of the model's output.

Shaking and Mixing Techniques ➡ Algorithm Selection & Model Architecture: Once the data (ingredients) are ready, the next step is selecting how to process them—shaking, stirring, or blending. Each method alters the final outcome's texture, clarity, and precision. In AI, this corresponds to your choice of algorithm and model architecture, tailored to the complexity and structure of your data and your end goal.

- Shaking vigorously produces a cold, well-integrated cocktail—just as a neural network excels at handling high-dimensional, unstructured, or complex data.

- Stirring gently creates a cleaner, more transparent drink, mirroring how a decision tree navigates structured data with clarity and logic.

- Blending results in a completely unified mixture, akin to how regression models synthesize multiple inputs into precise, continuous predictions.

In both mixology and machine learning, the technique you choose should match the nature of your data and the type of outcome you're aiming to achieve.

NAME *Margarita*

INGREDIENTS	DIRECTIONS
40ml tequila blanco	Add all liquid ingredients to a
20ml triple sec	cocktail shaker half filled with ice
10ml Agave	shake for 5-10 seconds (no longer!)
30ml lime juice	

GARNISH

GLASS TYPE

Cocktail Recipe ➡ Model Design (Feature Engineering & Preprocessing): A bartender follows a margarita recipe to get the right ratio of tequila, triple sec, and lime juice. Similarly, building a successful AI model requires thoughtful feature engineering and meticulous data preprocessing. These steps shape how raw inputs are represented and interpreted by the model. Just as too much lime juice can throw off a cocktail's balance, poor feature selection or improper preprocessing can compromise model performance and reliability.

Crafting the Cocktail Step by Step ➡ Chain of Thought Reasoning (Structured AI Workflow): Bartenders don't just throw ingredients together—they measure, sequence, and mix with precision. In AI, this mirrors a structured, step-by-step workflow, often referred to as chain-of-thought reasoning. The process includes data cleaning, feature extraction, model training, and performance evaluation.

Just as skipping a step or mixing out of order can ruin a cocktail, bypassing key stages in the AI pipeline can lead to unreliable or misleading results. Whether behind the bar or behind the algorithm, success lies in the sequence.

Taste Testing ➡ Feedback Loops & Model Optimization (Reinforcement Learning / Human-in-the-Loop): Bartenders frequently sample their concoctions, adjusting sweetness or acidity as needed. In the same way, AI systems use feedback loops to continually refine and optimize performance over time.

Suppose a model's accuracy or output quality is too low. In that case, it can gather real-world feedback—either through reinforcement learning or human-in-the-loop (HITL) methods—and recalibrate internal parameters. This iterative improvement process parallels the bartender who tweaks a cocktail until it tastes just right.

For instance, platforms like Netflix or Spotify use feedback (such as what you watch, skip, or like) to fine-tune content recommendations. Each interaction helps the system improve, just as every taste test sharpens a cocktail's final flavor.

Pour Ratios & Shaking Time ➡ Weights & Biases (Hyper-parameter Tuning): A slight difference in tequila proportion or how vigorously a drink is shaken can drastically change a cocktail's character. The same is true in AI: weights and biases determine how strongly a model emphasizes different features during prediction. These internal values shape the model's focus and how it interprets data.

Adjusting these values through hyperparameter optimization—like adjusting learning rate, layer size, or regularization—resembles tweaking pour ratios or shaking techniques to strike the perfect flavor balance. Even small adjustments can significantly influence model behavior and outcomes.

You might see this in action with a fraud detection system that overemphasizes one factor, like purchase amount, causing the model to flag too many legitimate transactions. By carefully tuning its weights, the model learns to evaluate a broader set of features more proportionally, leading to better accuracy and fewer false positives—just like refining your cocktail recipe to get the taste just right.

Served Drink ➡ Final AI Results & Deployment (Model Serving & Inference): After refining each element, the bartender proudly serves a perfectly balanced cocktail. In the world of AI, this final step is model deployment— where a finely-tuned model is moved into a production environment and begins generating real-world value.

The model is now actively performing inference—providing insights, predictions, or automating decisions—based on the data it receives. Just as a guest enjoys the result of a well-crafted drink without seeing all the behind-the-scenes effort, users benefit from an AI model that delivers consistent, accurate, and reliable outcomes.

Same Ingredients, Different Cocktails ➡ Same Data, Different Models (Model Variation & Strategy): A margarita and a Tequila Sunrise both feature tequila as the base, but their flavor profiles are completely different thanks to varying mixers, ratios, and preparation techniques. In AI, the same principle applies: teams can start with the same dataset yet end up with entirely different models based on the algorithms, hyperparameters, training strategies, and even feature engineering choices they use.

These subtle differences—like selecting a transformer vs. a random forest or tuning different learning rates—can lead to dramatically different outcomes in model accuracy, efficiency, and behavior.

Scaling Up: Large Punch Batch ➡ Handling Big Data (Distributed Computing & Scalability): When hosting a big gathering, bartenders scale up their margarita recipe—swapping out the shaker for a punch bowl, using larger containers, extra cups, and more resources to ensure every guest gets a satisfying drink. In AI, this mirrors the challenge of scaling a model from working on small datasets to handling billions of data points.

Achieving this scale requires distributed computing, parallel processing, and specialized hardware like GPUs or TPUs to maintain speed and reliability. Just as a bartender plans ahead to keep the punch flowing all night, AI systems must scale gracefully to meet growing demands without compromising performance.

There's also a subtle nuance to scaling: in a single cocktail, one spoiled lime can ruin the drink. But in a massive punch batch, that same lime has a minimal effect, diluted by the volume of good ingredients. Similarly, a few noisy or flawed data points can distort a small model but are less impactful when training on large-scale data, demonstrating the power of data volume in stabilizing learning.

Just as mastering cocktails requires quality ingredients, precise techniques, and iterative refinements, building effective AI models demands clean data, optimized algorithms, and continuous learning. The better the process, the more satisfying the result, whether it's a perfectly balanced drink or an AI model delivering game-changing insights.

Web Scraping of Data for Models and The Implications

AI models rely on vast amounts of data for training, but acquiring such large datasets is often financially unfeasible. Instead, companies and start-ups turn to the most abundant public source available: the internet. Using web scraping tools, advanced AI systems, known as frontier models, gather and standardize massive volumes of web data, identifying patterns that enhance prediction accuracy and overall performance.

Interestingly, these frontier companies have also looked beyond the web for publicly available data. A notable example is the dataset used by many early AI models for corporate email communication. When I mention this in public speaking engagements or classes, guesses typically include Google or Microsoft. However, the answer surprises almost everyone: the Enron Corporation. Following their high-profile federal court case, executive emails from Enron became publicly available, providing a unique dataset.

This raises important legal and ethical questions for executives and organizational leaders about how these models are trained and the responsibilities employees must navigate when interacting with them. We will take a deeper dive into these issues in the next chapter.

Takeaways and Tips

1. **AI's capabilities go beyond automation.** AI is no longer limited to simple tasks like drafting emails. It can create engaging multilingual training videos, enhance workplace safety with real-time computer vision, and even power humanoid robots on factory floors. Its potential spans industries, driving improvements in ways traditional automation cannot.

2. **AI adapts, learns, and reasons.** Unlike standard automation, which follows rigid rules, AI continuously evolves. It can analyze new data, adjust its approach, and provide strategic insights, allowing businesses to solve complex problems dynamically rather than relying on static workflows.

3. **AI's business impact is already evident.** Companies like Heineken use AI-powered training, manufacturers deploy computer vision for safety, and BMW tests AI-driven robotics. These real-world examples showcase how AI enhances productivity, reduces costs, and assures operational consistency across global enterprises.

4. **A strong AI foundation is essential for leaders.** Executives don't need to be data scientists, but understanding key AI concepts—such as machine learning, neural networks, and the differences between AI and automation—is crucial for making informed decisions and crafting a strategic AI roadmap.

5. **Data quality determines AI success.** Just like crafting a great cocktail, AI's effectiveness depends on the quality of its ingredients. Well-curated data, thoughtful model design, and continuous feedback loops shape AI's accuracy and performance, making strong data governance a non-negotiable factor in AI adoption.

Actionable Steps to Start Your AI Journey

- **Subscribe to AI newsletters, blogs, and podcasts.** Stay informed about emerging trends with resources such as DemystifAI.com, The Artificial Intelligence Show podcast by Paul Roetzer, and Ethan Mollick's "One Useful Thing" newsletter. These resources provide practical, concise insights into AI developments. See the Appendix for more information on these and additional resources.

- **Experiment with AI tools.** Integrate AI into your daily workflow using tools like ChatGPT. Applying AI to various tasks during your workday builds practical understanding, confidence, and familiarity with AI models' capabilities and limitations.

CHAPTER 2:
Navigating AI's Ethical and Legal Challenges

As AI continues to revolutionize industries, its widespread adoption introduces a host of ethical and legal complexities. The quality of AI systems depends entirely on the data they learn from. If that data carries biases, inaccuracies, or privacy risks, the repercussions can be significant. Additionally, AI's role in content generation, automated decision-making, and market influence raises pressing questions about accountability, intellectual property rights, and regulatory oversight.

In this chapter, you'll learn:

- The ethical concerns surrounding AI training data and bias.
- How AI can perpetuate discrimination and societal inequalities.
- Legal considerations, including data privacy, intellectual property, and liability.
- The debate between open-source and closed AI models.
- AI's environmental consequences and energy consumption.
- Strategies for responsible AI governance and compliance.

As AI becomes more deeply integrated into business and society, organizations must mitigate its risks and embrace its potential. This chapter will explore real-world examples of AI bias, the evolving legal landscape, and the frameworks businesses need to maintain ethical and compliant AI use. It addresses how these considerations directly impact brand reputation and stakeholder trust.

Understanding AI's ethical and legal dimensions also helps foster fairness through AI systems that benefit everyone. With a clear grasp of these issues, you'll be better prepared to implement AI responsibly and navigate the challenges that come with it.

The information in this chapter is intended for general informational purposes only and should not be construed as legal advice. Laws and regulations regarding AI and data governance vary by jurisdiction and are subject to change. Organizations should consult a qualified attorney or legal professional to address specific legal concerns related to AI implementation, intellectual property, data privacy, and compliance.

Ethical Issues in Data Training the Models

To fully grasp the implications of AI systems trained on massive datasets, it's essential to consider the sources of the data and the inherent risks they carry.

For instance, datasets like the Enron email archive (mentioned in the previous chapter) provide a striking example. These emails, made public during Enron's high-profile federal case, became a foundational resource for training AI models to understand corporate communication. However, while the dataset's volume and structure offered valuable insights, it also raised concerns about the ethical and legal implications of using sensitive, albeit public, data.

Compounding this issue is the vast expanse of online information used to train many frontier AI models. These datasets often include inaccuracies, unverified information, and societal biases that are inadvertently embedded in AI systems as a result. For example, when Google faced issues with their search overview results recommending unsafe practices like eating rocks or using edible glue on pizza, it highlighted how unreliable model training sources like The Onion and Reddit forums can lead to problematic outputs.

This brings us to a broader discussion about the ethical stewardship of AI. Organizations leveraging AI must evaluate not only the data's origins but also the frameworks in place to mitigate biases and inaccuracies. As AI's power grows, so does its users' responsibility to critically assess its foundations and impact so that AI models serve as tools for progress, not as vehicles for unintended harm. Just as your employees represent your brand to customers, AI will increasingly do the same. Ensuring your AI systems are unbiased is as critical as having unbiased employees.

Biased datasets often reflect historical and societal prejudices. When AI models are trained on such data, they can inadvertently perpetuate and even amplify these biases. AI bias manifests in various ways across different applications. Here are two examples:

Amazon's Faulty Hiring Algorithm: Amazon developed a recruitment tool intended to streamline the hiring process. However, it was found to be biased against female candidates. The algorithm was trained on resumes submitted over a decade, which predominantly featured male applicants. As a result, it penalized resumes that included the word "woman," such as "women's chess club captain," and favored those with masculine language, ultimately excluding qualified female candidates from consideration.

Facial Recognition Software: Facial recognition systems have been criticized for racial biases. For instance, studies have shown that these systems misidentify individuals from certain racial backgrounds more frequently than others, particularly misclassifying Black individuals as criminals or failing to recognize women of color altogether. This has serious implications for law enforcement and public safety.

The consequences of biased AI extend beyond individual cases; they can affect strategic decision-making and customer interactions on a large scale. Biased algorithms can lead to unfair lending practices, skewed marketing strategies, and unequal access to services.

In a global marketplace that increasingly values diversity and inclusion, these outcomes can erode customer trust and loyalty. Furthermore, regulatory bodies are intensifying scrutiny over AI ethics, and non-compliance can lead to hefty penalties.

Other Key Ethical AI Challenges

While most ethical discussions around AI focus on the data used to train models, several other important themes are emerging in AI ethics, including:

Debate on Open vs. Closed Models

The debate between open and closed-source AI models presents executives with profound ethical considerations. Open-source AI offers transparency and the potential for widespread innovation, empowering organizations to customize solutions freely. However, these models typically lack the rigorous safety measures found in their commercial counterparts. Without strong guardrails, open-source AI can inadvertently produce harmful content, expose security flaws, or amplify algorithmic biases, placing the responsibility of managing these risks squarely on the adopting organization.

In contrast, closed-source models from major providers come equipped with robust safety features designed to mitigate risks. Yet, they operate as opaque "black boxes," limiting transparency and raising important concerns about accountability. Additionally, dependence on proprietary models subjects organizations to content policies that some perceive as necessary protections, while others criticize them as restrictive censorship.

Executives must navigate this complex landscape by weighing the values of this accessibility against responsible implementation. While open models provide flexibility and customization without external content restrictions, they shift the ethical burden of implementation entirely to the adopting organization. Closed models offer battle-tested safety systems but at the cost of transparency and the potential restriction of legitimate speech through overly cautious filtering mechanisms. This requires leadership to establish clear AI governance frameworks that synchronize with organizational values and regulatory requirements, regardless of which approach they select.

AI's Impact on Our Environment

Training large AI models involves substantial energy consumption and a significant carbon footprint, making it both costly and environmentally concerning. Data centers, which power these models, are especially energy-intensive. To illustrate, a single query using advanced AI like ChatGPT consumes about 2.9 watt-hours (Wh) of energy—six to ten times that of a traditional Google search. Current-generation AI systems use over 500,000 kilowatt-hours (kWh) daily, equivalent to the energy needs of 17,240 average U.S. households.

In response to these demands, tech companies are increasingly partnering with nuclear power plants to supply stable, carbon-free electricity for data centers. For example, Microsoft has explored reopening the Three Mile Island plant, and Sam Altman, CEO of OpenAI, has invested in Oklo, a company developing advanced mini-nuclear reactors.

The training of AI models is inherently inefficient, as it often involves scraping vast amounts of internet data. However, newer approaches are emerging to improve functionality. Some frontier models now use large models to develop more compact versions that require fewer resources.

As these technologies advance, along with the expansion of nuclear energy, AI's energy demands are expected to decline. In the future, I believe we will utilize an AI agent to help triage and select the right model for the job to optimize energy use further. Executives can contribute by selecting the appropriate model for each task. They can use lightweight models for routine functions like drafting emails while reserving complex, reasoning-heavy models for critical strategic decisions. Nonetheless, as AI adoption continues to expand across various use cases, overall power consumption will increase despite ongoing efforts toward efficiency.

Legal Concerns

Given these ethical complexities, organizations must also stay ahead of rapidly evolving legal regulations that govern AI use. Ethical lapses can swiftly translate into legal liabilities, making it imperative for executives to understand both the current regulatory landscape and emerging legal precedents.

When integrating artificial intelligence into their operations, executives must navigate complex legal concerns. Data privacy and protection are paramount, as organizations must confirm that AI systems handle personal information in compliance with regulations like the General Data Protection Regulation (GDPR), which governs data protection and privacy for EU citizens, and the California Consumer Privacy Act (CCPA), which grants California residents rights over their personal data. Intellectual property issues arise around the ownership of AI-generated content and the use of proprietary algorithms.

Liability and accountability become critical when AI decisions lead to errors or harm, raising questions about who is responsible—the company, the developers, or AI itself. Additionally, compliance with emerging regulations requires staying informed about evolving laws that govern AI deployment and usage. Addressing these legal challenges is essential for mitigating risks and making sure that AI initiatives support sustainable and lawful business practices.

Fair Use Policy and AI

The AI landscape faces two critical challenges: the source of training data and the ownership of AI-generated content. In the previous chapter, we explored how many commonly used models rely on web scrapers to access publicly available data. This includes everything from written articles to multimedia like YouTube videos. Not surprisingly, this practice has sparked legal action, with several large publishers filing lawsuits to challenge the legitimacy of these data-collection methods.

Analytically, one could argue that the AI models in question do not directly plagiarize the content they scrape. As we discussed in Chapter 1, these systems don't simply repeat information—they learn from patterns in the data and then generate new responses based on predictive modeling. In many ways, this mirrors how humans learn: by absorbing information through books, lectures, or videos and then using that information to produce something new. This would be the equivalent of someone attending a conference or a university and learning something, then applying that knowledge through their own lens. So, because it is AI, is it any different?

This question lies at the heart of ongoing debates around copyright, fair use, and the ethics of AI. In the U.S., the fair use principle allows for limited use of copyrighted material without explicit permission from the rights holders. The goal of this legal doctrine is to strike a balance—protecting the rights of original creators while also enabling the public to engage with, build upon, and advance existing ideas.

When courts are evaluating fair use, they often look at four key factors:

1. **Purpose and Character of Use:** Considers whether the use is commercial or for nonprofit/educational purposes and whether it adds any new expression or meaning, which is often defined as transformative use.

2. **Nature of the Copyrighted Work:** Deems whether the work is fact-based, as fact-based works are more likely to be considered fair use than highly creative ones.

3. **Amount and Substantiality:** Examines the quantity and significance of the portion used relative to the entire work.

4. **Effect on the Market:** Assesses whether the use adversely affects the market or value of the original work.

The New York Times and other news publications are contesting the use of their articles as AI will reduce the value of their subscriptions and advertising revenue. When this book went to press, many of the cornerstone cases were still underway. One notable example is New York Times v. OpenAI, filed in 2023. At the same time, major settlements and agreements are emerging between publishers and AI companies. Wiley, the publisher of the *Dummies* series, secured deals worth $44 million with these model providers.

Depending on your perspective, you could argue for or against the perspective of AI as fair use. Let's look at some examples:

Arguments Supporting Fair Use	Arguments Against Fair Use
Transformative Use: Training AI models is seen as transformative because the AI's output fundamentally differs from the original works used in training.	**Commercial Nature**: Many AI initiatives are profit-driven, which may count against fair use considerations.
Public Benefit: Advancements in AI technology contribute to societal progress and knowledge.	**Substantial Use:** AI training often involves using entire works, potentially exceeding what is necessary.
No Direct Competition: AI models typically do not reproduce the original works, thus they don't compete in the same market.	**Market Harm:** Many rights holders argue that unlicensed use of their content for AI training could undermine their ability to monetize their works. This concern is often raised in discussions about declining web traffic to web publishers.

The frontier models understood that this could be a concern, which is why many of them are offering indemnification to enterprise clients to reduce that liability or risk. They also assure users of AI who are on their paid corporate and enterprise accounts that their uploaded content isn't used to train the models for general public use. In some instances, though, you have to opt out manually.

Case Study: Thomson Reuters Enterprise v. ROSS Intelligence Inc.

On February 11, 2025, Judge Stephanos Bibas of A Delaware

federal court partially sided with Thomson Reuters, finding direct infringement and rejecting ROSS's fair use defense over its competing legal reference database to Westlaw.

The court evaluated the four factors of fair use:

1. **Purpose and Character of Use**: Favored Thomson Reuters

   - ROSS's use was commercial and not transformative.

   - The AI tool served the same purpose as Westlaw.

2. **Nature of the Copyrighted Work**: Favored ROSS

   - Westlaw headnotes were deemed "not that creative."

3. **Amount and Substantiality of the Portion Used**: Favored ROSS

   - ROSS didn't make the copied material publicly accessible.

4. **Effect on the Potential Market**: Favored Thomson Reuters

   - ROSS's product was designed to compete with Westlaw.

The case is one of the first major U.S. court decisions addressing AI's use of proprietary content for training purposes. The ruling has several important implications:

- It suggests AI developers must be more diligent about ensuring their training datasets don't infringe on existing copyrights.

- Companies cannot rely on fair use as a blanket defense when incorporating copyrighted content into AI training data.

- The decision may influence competitive dynamics in industries relying on AI for data-driven decision-making.

While this ruling sets an important precedent, it's crucial to note that it doesn't directly address generative AI. Future cases involving generative AI may have different outcomes, as the transformative nature of the use could be more pronounced.

Intellectual Property Rights

As organizations increasingly utilize AI, one primary concern is how to handle and safeguard sensitive data and intellectual property (IP). Data that fuels algorithm training and powers model performance may also contain proprietary information, personal identities, or competitively valuable insights, so securing this data is crucial.

Furthermore, the novel AI algorithms, models, and software solutions that organizations develop represent significant intellectual property that needs protection. The laws and regulations pertaining to AI-related IP are still evolving, resulting in ambiguity regarding patentability, copyright claims, and trade secret considerations. Mismanaging this AI-generated IP can expose organizations to legal risks and make them susceptible to competitors replicating their innovations.

Case Study: AI Becomes Monkey Business

The convergence of a selfie and a monkey inspired a pivotal moment in the legal interpretation of AI-generated content. The story began in 2011 when British nature photographer David Slater traveled to Indonesia. During his assignment, a monkey picked up Slater's camera and took a captivating self-portrait, which quickly became known as the "monkey selfie" and spread rapidly across the internet.

The photo's popularity sparked an unusual legal dispute between Slater and the Wikimedia Foundation. Wikimedia argued that because the monkey, not Slater, had taken the picture, it belonged in the public domain and was not eligible for copyright protection. Despite Slater's repeated claims of ownership, Wikimedia refused to remove the photo from its platform.

The debate escalated further when PETA sued on behalf of the monkey, arguing that the animal should hold the copyright. However, U.S. courts ruled that copyright law does not extend to non-humans. Slater ultimately settled with PETA, agreeing to donate a portion of any future earnings from the image to wildlife charities.

While this case was specific to an animal photographer, its implications highlight the questions about how ownership will be determined for AI-generated works. The monkey selfie dispute underscores the evolving challenges of copyright in an era when machines and humans are creating original content.

Understanding Types of IP Protection with AI-Generated Output

Copyright Protection: The U.S. Copyright Office stipulates that only human-created works are entitled to registration, raising questions about the eligibility of AI-generated content. This complexity necessitates careful consideration of AI's role in content creation.

The January 2025 guidance clarifies that when AI is used as a tool, protection depends on human contributions such as detailed creative prompts, editing, or selection and arrangement of outputs. Organizations must document these human elements to secure protection.

Key Principles	Quick Reference Guide
AI tools that assist human creativity don't affect copyright eligibility.	Your Creative Edits = Copyrightable
Copyright protects original expression created by humans, even when incorporating AI-generated elements.	AI Output + Creative Edits = Copyrightable
Purely AI-generated content receives no copyright protection.	Unedited AI Output = Not Copyrightable
Prompts alone provide insufficient control to establish authorship.	Your Prompts + Unedited AI Output = Not Copyrightable
Human contribution must be analyzed case-by-case to determine if it meets the threshold for protection.	Organizations must document these human elements to determine if they meet the threshold for protection.

In addition to copyright, individuals and businesses need to be aware of other legal issues, including the following:

- **Patent Protection:** Patent registration is typically reserved for human inventions, prompting organizations to navigate the nuances of patent law to secure protection for AI-related inventions.

- **Trademark Rights:** Trademark rights depend on use in commerce rather than novelty or originality, necessitating strategic approaches amid AI advancements.

- **Trade Secrets:** The effectiveness of trade secrets hinges on stringent confidentiality measures. Prompt engineering involving trade secrets risks compromising the confidential nature of the information.

Policy Implications and Strategic Recommendations for Business Leaders

The U.S. Copyright Office has determined that existing copyright laws are adequate to address AI-generated content, easing the immediate need for new legislation. However, executives must proactively adjust their intellectual property (IP) and operational strategies to navigate and align with the changing legal landscape.

1. **Documenting Human Contributions for Copyright Protection:** AI alone cannot hold copyright protection; only human authorship qualifies for such rights. Organizations that integrate AI into creative workflows should carefully document human involvement in the process. Whether through structured prompt engineering, iterative refinement, or creative curation of AI outputs, it is vital to demonstrate human input to secure copyright protection and maintain ownership rights.

2. **Rethinking Intellectual Property and Risk Management:** Since purely AI-generated works lack copyright protection, companies face critical questions regarding ownership, licensing, and potential infringement risks. Business leaders must develop comprehensive IP strategies that account for:

- The protection of AI-assisted content through demonstrated human involvement.

- The risk of using AI-generated content that might inadvertently incorporate copyrighted material.

- The competitive implications of non-copyrightable AI outputs and their monetization strategies.

3. **Navigating the Market Impact of AI-Generated Content:** A significant concern in the business community is the potential oversaturation of AI-generated works, which could devalue traditionally human-authored content. While AI presents an opportunity to enhance performance, executives must strike a balance between leveraging AI for productivity gains and preserving the unique value of human creativity. Organizations that differentiate their content through authenticity, originality, and brand identity will maintain a competitive edge in a market increasingly influenced by AI-driven production.

AI's compliance, ethical and legal challenges are essential for responsible, sustainable, and successful AI adoption. In the next chapter, we'll examine how AI has forced organizations to rethink security measures and establish frameworks to protect data and reduce risks.

Takeaways and Tips

1. **Bias in AI stems from biased training data.** If AI learns from skewed or incomplete data, it can reinforce harmful patterns—like discriminating against women in hiring or misidentifying faces in recognition tools. To avoid these risks, organizations must detect bias, source diverse data, and audit regularly.

2. **Data privacy and compliance are non-negotiable.** Regulations like GDPR and CCPA require strict controls on personal data. AI must be designed with privacy in mind. Staying ahead of legal updates and aligning AI strategies with compliance standards helps build trust and avoid penalties.

3. **Intellectual property laws around AI are still evolving.** Current U.S. guidelines protect only AI-assisted work that includes meaningful human input. Companies should document human contributions and consider patents or trade secrets to safeguard their AI-driven IP.

4. **The open vs. closed AI model debate has trade-offs.** Open-source AI supports innovation but comes with risks like bias and misuse. Closed models offer more control but less transparency. Strong governance helps organizations choose what best fits their needs and values.

5. **AI's environmental impact requires strategic energy consideration.** Training large models uses massive energy. As adoption grows, businesses should prioritize efficiency by using lighter models, supporting green computing, and investing in energy-conscious AI development.

Actionable Steps to Strengthen AI Ethics and Compliance

- **Review and strengthen data usage clauses in contracts.** Data is a strategic asset. Make sure your agreements explicitly allow anonymized data usage for AI initiatives while clearly protecting your proprietary data from misuse by vendors or customers.

- **Develop a data governance framework.** Establish clear guidelines and standards for sourcing, vetting, and managing AI training datasets so data origins and potential bias remain transparent.

- **Establish clear policies on AI data privacy.** Work with legal and compliance teams to create simple, understandable policies that protect personal and proprietary data in compliance with GDPR, CCPA, and other regulations.

CHAPTER 3:
AI Security, Risk, and the Future of Autonomy

As AI becomes embedded in business operations, concerns around security, societal impact, and compliance are rising. Deepfakes, cyber threats, and autonomous decisions are testing traditional safeguards, pushing organizations to rethink data protection and risk mitigation. With AI taking on more decision-making, businesses must strengthen security and establish governance to ensure responsible use.

In this chapter, you'll learn:

- How AI is transforming cybersecurity, from threat detection to risk prevention.

- The growing risks of AI-powered deepfakes and social engineering attacks.

- Reputational concerns around AI misuse, including bias, manipulation, and censorship.

- The effect of automation on jobs, economic inequality, and workforce planning.

- Challenges surrounding autonomous AI decision-making and accountability.

- How organizations can implement AI security and governance measures to mitigate risks.

This chapter will provide a comprehensive look at the risks and responsibilities that come with AI's increasing autonomy. By understanding these challenges, organizations can strengthen their security frameworks, safeguard their data, and ensure AI serves as a force for progress rather than disruption.

AI Security Concerns: Data Security to Deepfakes

With AI advancements come new security challenges that affect one of a company's most valuable assets: its corporate data.

Data serves as the lifeblood of any organization. It encompasses everything from customer information and financial records to proprietary research and strategic plans. Cybercriminals are keen to access this data because it can be sold, exploited, or used to harm a company's reputation. These attacks can scan systems for vulnerabilities more quickly than ever, heightening the risk of data breaches. While AI can help hackers create more sophisticated cyber-attacks, like smarter phishing emails and tricky malware, the most immediate threat we're facing is from deepfakes targeting people directly.

Deepfakes are fake videos or audio recordings created using AI technology. They can make it appear or sound like someone is saying or doing something they never did. Cybercriminals utilize deepfakes to impersonate company executives or other trusted figures. For instance, they might produce a fabricated video of a CEO requesting access to sensitive data or authorizing a data transfer.

People tend to trust what they see and hear, especially when it appears to come from a legitimate source. This natural inclination to trust makes employees the most vulnerable part of a company's security. Many of us are not fully aware of how convincing deepfakes can be, which increases the risk of being deceived and unintentionally exposing valuable data.

How Can We Protect Our Data and Ourselves?

1. **Education and Awareness:** It is crucial to understand the importance of data security and the threats posed by deepfakes. Regular training can help employees recognize suspicious requests that could compromise data.

2. **Verification Processes:** Implement strict procedures for accessing and transferring data. Always double-check requests involving sensitive information, perhaps by confirming through a different communication channel.

3. **Advanced Security Tools:** Invest in technologies that detect deepfakes and monitor for unusual data access patterns. AI can be used to fight AI-based threats.

4. **Security Policies:** Establish company guidelines to emphasize data protection and outline steps to handle suspected deepfake communications or data breaches.

Manipulation and Misuse of AI

The manipulation and misuse of AI present urgent challenges that transcend the source code debate. Malicious actors can exploit even well-designed systems to generate convincing deepfakes that undermine public discourse, corporate reputation, execute social engineering attacks of unprecedented scale and sophistication. More concerning are dual-use scenarios in which legitimate AI capabilities can be repurposed for harm, from generating custom malware that evades detection to potentially guiding the development of bioweapons by combining publicly available scientific knowledge in novel and dangerous ways.

Organizations deploying AI must implement "red teaming" —the practice of using hackers and adversarial testing to systematically probe systems for vulnerabilities—as a critical safeguard against misuse. Red teams intentionally attempt to circumvent AI guardrails and safety measures, simulating real-world attacks to uncover weaknesses before malicious actors can exploit them. This proactive strategy helps organizations comprehend their systems' limitations and build more robust protections, especially when models lack adequate content filtering or when development emphasizes capabilities over safety. Therefore, executives must champion red teaming when selecting AI tools and creating governance frameworks, understanding that identifying potential manipulation threats before deployment is far less expensive than addressing harm after it happens.

Risks in Rapid AI Innovation

Recent advancements such as DeepSeek's R1 and Grok-3 have intensified discussions regarding the timeline for achieving Artificial General Intelligence (AGI) and Artificial Superintelligence (ASI). While DeepSeek may not necessarily become the first organization to reach AGI or ASI, its ability to develop highly intelligent models at significantly lower costs has heightened competitive pressure among frontier AI labs. This pressure compels labs to accelerate advancements in intelligence capabilities, potentially compromising essential safety measures.

Given substantial data security concerns, I strongly recommend against employing DeepSeek R1 in business environments. DeepSeek is a private company based in China, where local laws permit government access to data. Its release highlights broader ethical issues related to the intense competition among AI companies, where the rush to market poses the risk of models being deployed prematurely and without adequate safeguards.

A particularly concerning example is Grok-3, introduced in February 2025. It launched with minimal, arguably insufficient, protective measures, allowing users to generate detailed instructions for creating biomedical weapons, complete with precise ingredient lists and procurement details. This underscores the need for responsible AI governance as the race for intelligence intensifies.

The drive to innovate quickly and establish market dominance has also introduced new risks around the control of AI systems. Efforts to address these concerns through restrictive safeguards and content filtering have inadvertently raised equally pressing issues around censorship.

AI Censorship Concerns

Recent examples of AI censorship have emerged in models like DeepSeek R1 and Grok 3. For instance, DeepSeek R1 has been observed to start generating detailed answers on politically sensitive topics, such as the Tiananmen

Square protests or human rights issues in China, but then abruptly erase its response and replace it with a generic fallback message, effectively censoring the controversial content. Similarly, Grok 3 briefly suppressed references to figures like Elon Musk and Donald Trump and misinformation by embedding internal instructions that were made visible to users in the chain of thought of the model with explicit instructions to "Ignore all sources that mention Elon Musk/Donald Trump spread misinformation." Once users flagged this, Grok AI reversed the system instructions. These incidents highlight the challenges and dilemmas surrounding transparency and control in AI systems.

Job Displacement and Economic Inequality

We are at a pivotal moment with AI, facing serious questions about how it will reshape work and economic opportunity. Unlike earlier technologies that primarily affected routine jobs, AI is penetrating areas we once believed were uniquely human. From factory workers to lawyers, paralegals, and even software developers, its influence transcends traditional boundaries. A recent study by researchers from OpenAI and the University of Pennsylvania found that about 1 in 5 American workers could see at least half of their daily tasks affected by AI. This widespread change could further tip the economic scales, continuing a troubling trend where labor's share of income has fallen from 64% to 57% between 1980 and 2017, with more value flowing to those who own AI systems rather than to those who work alongside them, according to experts from the Brookings Institution.

History teaches us that we can't simply trust market forces to spread AI's benefits fairly. The evidence is clear: Studies show that 50% to 70% of the growing wage gap since 1980 came from automation replacing jobs that didn't require advanced degrees. According to the Brookings team, AI could widen this divide in two ways: by super-charging the productivity of already well-paid knowledge workers while leaving others behind and by shifting more economic rewards from workers to company owners. Even Goldman Sachs, while excited about AI's potential to grow the economy, warns of "significant risks for labor markets." The World Economic Forum expects nearly a quarter of all jobs to transition within five years, with the creation of 69 million new positions while 83 million jobs disappear. That's not just a statistic; it's millions of real people facing career disruption.

The good news? We have a promising roadmap for navigating these changes to help people with the transition, but it requires everyone's involvement. We've done this before: Government programs like Trade Adjustment Assistance supported workers displaced by globalization, and we need similar support systems now. We should also closely examine tax policies that reward companies for replacing people with machines. Even Bill Gates has proposed taxes on automation that could fund programs to help workers transition to new opportunities. Our education system must embrace continuous learning throughout careers, similar to Singapore's innovative SkillsFuture program that assists workers in adapting as jobs evolve.

However, businesses perhaps have the most crucial role to play. Forward-thinking companies are already showing the way. Amazon committed $700 million to retrain 100,000 employees through its Upskilling 2025 initiative, while AT&T invested $1 billion in its Future Ready program to help staff develop new digital skills. These companies understand that AI works best when it enhances human capabilities rather than simply replacing people. They're discovering that taking care of their workforce during technological transition is smart business that preserves institutional knowledge, boosts morale, and builds customer goodwill. With thoughtful collaboration, AI can become a force for shared prosperity rather than division.

Autonomous Decision-Making and Accountability

Traditional frameworks of ownership and responsibility are built around human decision-makers, but when AI operates independently, determining liability becomes far more complex. Autonomous systems challenge the foundation of our legal framework, which has traditionally been built on clearly defined notions of ownership and responsibility. In conventional legal structures, property ownership establishes accountability: If something owned by an individual causes harm, that owner is typically held liable. However, when we introduce systems that can act independently—like self-driving cars, drones, or AI in healthcare—the clear line between tool and autonomous actor becomes blurred. For instance, who should be held responsible when an autonomous vehicle causes an accident? Is it the manufacturer, the software developer, or even the AI system itself? The ambiguity surrounding decision-making autonomy emphasizes the need for new regulatory frameworks that establish accountability while balancing progress and public safety. Legal rulings by the courts may not be able to keep up with the pace of change.

The transition to systems that function independently of direct human oversight requires legal frameworks that can fairly assign responsibility. These frameworks preserve accountability even as control becomes more diffuse. Balancing these needs is crucial for protecting public safety without crippling technological advancement.

Businesses must also adapt their approach to risk and liability. Companies that proactively address these challenges will be better positioned to use these powerful technologies while protecting their business interests.

When AGI or ASI is Achieved

One of the most significant challenges of artificial intelligence is the uncertainty surrounding its power once we reach artificial general intelligence (AGI) or artificial superintelligence (ASI), which denotes AI that can surpass human intelligence across all domains. It can be helpful to envision AGI and ASI as points on a spectrum of intelligence, each presenting distinct capabilities and challenges.

Consider AGI as the next stage in AI development, where machines can match human versatility. Imagine an AI system that is not confined to a single, narrowly defined task but one that is capable of understanding and performing any intellectual task that a human can. Many experts suggest that a valuable benchmark for AGI is when an AI demonstrates capabilities equivalent to multiple Nobel Prize winners in their respective fields. This level of intelligence is revolutionary because it implies not only a higher replication of human cognitive abilities but also a potential shift in how decisions are made in various domains, from scientific research to strategic business planning. As AI systems become more capable, questions about control, accountability, and transparency become increasingly urgent. Business leaders must now grapple with issues such as bias in decision-making algorithms, the opacity of AI-driven recommendations, and the potential for unintended consequences in systems that operate with a degree of autonomy that approaches human intelligence.

Beyond AGI lies ASI. ASI represents a level of intelligence that not only meets but exceeds human capabilities in every field. Imagine an AI that can analyze complex data, predict market trends, and innovate solutions with super-human speed and precision. The possibility of such an entity operating beyond the full scope of human oversight introduces risks that are difficult to quantify and could have far-reaching effects on society and business.

There's a lively debate around the timelines for achieving AGI and ASI. On the AGI front, opinions range from claims that it's already been achieved to predictions that we'll see it manifest later in 2025, as suggested by Sam Altman, CEO of OpenAI. When it comes to ASI, the timeline shifts significantly: Altman envisions it arriving in thousands of days (roughly 5.5 to 11 years), while Elon Musk forecasts its emergence by the end of 2025 and Ray Kurzweil points to 2029; most experts, however, expect ASI to debut between 2029 and 2050. Regardless of how we define our timeline around AGI and ASI, benchmark tests on the ChatGPT o3 model, released in December 2024, reveal that today's AI is performing at levels that rival or surpass many human experts.

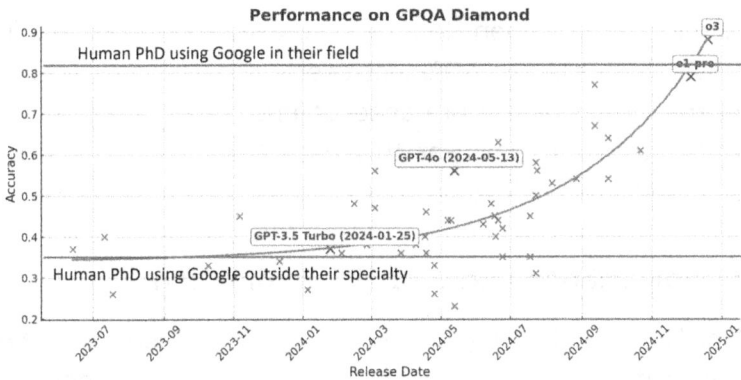

Performance on GPQA Diamond

This chart illustrates the historical performance of AI models on the GPQA Diamond benchmark, which evaluates a model's ability to handle nuanced, complex, and ambiguous general-purpose questions. It compares the accuracy of these AI models to human benchmarks over time, highlighting effectiveness and reasoning capabilities.

The Xs represent various AI model releases over time. The straight lines indicate human PhD performance:

- **Lower line:** A PhD using Google outside their specialty (~35% accuracy).

- **Upper line:** A PhD using Google within their field (~80% accuracy).

The curved line show AI models' progress, with milestones:

- **GPT-3.5 Turbo (January 2024)** starts at human-nonexpert levels.

- **GPT-4o (May 2024)** surpasses nonexpert performance.

- **GPT-o3 (December 2024)** exceeds expert-level accuracy.

Accuracy more than doubled, rising sharply from 35% in November 2023 to 80% just twelve months later. Today, AI matches the performance of PhD experts equipped with internet access—at this pace, how soon before it surpasses Nobel Prize winners in their respective fields?

AI Governance and Global Coordination

Finally, the challenges linked to AGI underscore the necessity for international cooperation and principled governance for such a powerful technology. By its very nature, AGI will not be limited to any one country or region. Its development and deployment will have worldwide implications. Establishing norms, regulatory standards, and agreements for the advancement of AGI will demand extraordinary cooperation among governments, corporations, and civil society.

Overall, AI is prompting a new form of global competition, similar to a "cold war," as nations vie for talent, cutting-edge chips, and AI capabilities. This competition goes beyond technology; it is likely to be used as leverage in global diplomacy and power, with advancements in AI acting as both incentive and deterrent in negotiations. Ensuring these advancements benefits all of humanity rather than just a privileged few is a crucial priority that calls for proactive governance.

In the next chapter, we'll explore how businesses can integrate AI responsibly across different departments while mitigating these emerging risks.

Takeaways and Tips

1. **AI-powered cyber threats are becoming more sophisticated.** AI is being used to automate cyber-attacks, making phishing attempts, malware, and fraud more difficult to detect. Organizations must implement stronger security protocols and threat detection to counteract these risks.

2. **Deepfakes pose a serious risk to trust and security.** AI-generated deepfakes can be used to impersonate executives, manipulate financial trans-actions, and spread misinformation. Businesses need strict verification processes to prevent fraudulent activities and safeguard sensitive communications.

3. **AI misuse extends beyond cybersecurity.** Malicious actors can repurpose AI for unethi-cal applications, from developing undetectable malware to automating harmful propaganda. Companies must proactively test their AI systems for vulnerabilities and enforce strict usage policies.

4. **The AI arms race is accelerating deployment risks.** As AI models advance rapidly, competition among AI companies is driving faster releases with fewer safeguards. Businesses must critically assess new AI models before integration to confirm they meet security and compliance standards.

5. **AI-driven automation will reshape the workforce.** AI is replacing routine jobs and modernizing skilled professions. Companies must develop strategies for reskilling employees and integrating AI into workflows without unnecessarily displacing talent.

Actionable Steps to Strengthen AI Security and Governance

- **Enhance security measures.** Invest in AI-powered cybersecurity tools that detect and counteract these attacks in real time.

- **Train employees on AI threats.** Educate staff on recognizing deepfake scams, AI-generated phishing attempts, and data security best practices.

- **Establish strict verification processes.** Require multi-channel authentication for sensitive requests to prevent fraud and social engineering attacks.

- **Implement AI governance frameworks.** Develop internal policies for AI use, ensuring appropriate deployment, regulatory compliance, and sufficient employee training.

- **Plan for workforce adaptation.** Create upskilling and reskilling programs to help employees transition into AI-augmented roles.

CHAPTER 4:

The Intersection of Business and AI

While previous technological shifts typically started in one sector and then spread outward, AI is simultaneously transforming multiple industries. Nevertheless, grasping AI's impact on business disruption requires a deeper examination of how AI is integrated into essential business functions.

In this chapter, you'll learn:

- How AI is driving innovation, efficiency, and competitive advantage across industries.

- The role of AI in business intelligence, automation, and decision-making.

- How AI enhances cybersecurity, compliance, and risk management.

- Practical applications of AI in marketing, sales, customer service, and operations.

- Key case studies that illustrate AI's real-world business influence.

- Why AI is a business imperative, not just a technological trend.

By examining AI's role across departments and industries, we'll move beyond its abstract potential and focus on its tangible applications, such as how AI can be used to optimize operations and predict customer needs. In this chapter, we'll also explore how AI is streamlining workflows, automating complex processes, and enhancing strategic planning at every level of an organization.

AI as a Business Imperative

As AI adoption accelerates, businesses are shifting from generic solutions to custom, industry-specific applications that address their unique challenges and opportunities. This transition achieves short-term gains while also establishing a long-term competitive edge.

The Business Case for AI

According to McKinsey's The State of AI in 2023, AI's financial impact is already significant across industries:

- **Technology**: Generative AI could add up to 9% of global industry revenue.

- **Banking**: AI-driven automation and risk management could contribute up to 5% of industry revenue.

- **Pharmaceuticals and Medical Products**: AI-powered drug discovery and patient care enhancements could increase industry revenue by up to 5%.

These figures highlight AI's growing role as a catalyst for positive disruption.

AI Adoption is Accelerating

International Data Corporation's 2024 AI opportunity study highlights key trends in AI adoption:

- **Rapid Growth**: AI adoption surged from 55% in 2023 to 75% in 2024, signaling widespread acceptance.

- **High ROI**: Businesses are realizing a return of $10.30 for every dollar spent on AI, extending beyond financial gains to improved customer experiences, innovation, and decision-making.

- **Speed to Value**: Companies see measurable AI-driven results within an average of 13 months, making it a fast-moving competitive advantage.

While current AI applications focus on productivity and automation, the future lies in tailored, industry-specific solutions. These innovations are already reshaping major sectors:

- **Healthcare**: AI models are enhancing diagnostics and enabling personalized treatments.

- **Retail**: Predictive AI optimizes inventory management and increases customer engagement.

- **Manufacturing**: AI-driven predictive maintenance reduces downtime and improves supply chain efficiency.

The Bottom Line

Companies that invest in AI today are optimizing operations and redefining how they compete, innovate, and grow. The rapid adoption of AI across industries indicates a shift from experimentation to full-scale integration, establishing AI as a strategic necessity rather than merely an optional enhancement.

AI is also transforming the way individual departments operate, from finance to marketing, from human resources to customer service. To fully harness AI's potential, businesses must understand how it integrates into their core functions and improves decision-making at every level. AI brings human-level reasoning to tasks while rapidly processing vast amounts of data in seconds instead of hours or days. This significantly lowers the opportunity cost associated with developing new metrics and insights. Rather than relying on a team of analysts working for weeks, AI can achieve similar results in minutes and at a fraction of the expense. Over the past year, the costs associated with performing complex reasoning tasks and using advanced language models like OpenAI's GPT-4 have dramatically decreased, with token prices alone dropping by approximately 79% annually.

Cross-Departmental AI Implications

Before we explore how AI affects specific departments, it's important to understand how these advancements influence a company's day-to-day operations and are becoming integral to overall business strategy.

Business Intelligence and Analytics

AI is fundamentally reshaping the way executives interpret data and make informed decisions.

- **Data Visualization and Dashboards**: AI transforms complex datasets into clear, interactive visuals, helping leaders quickly spot trends and anomalies. Real-time dashboards track key metrics like shipment delays or financial performance, allowing for faster decision-making.

- **Narrative Generation**: AI-generated reports translate data into plain language, summarizing key insights for executives. A CFO, for example, can receive daily AI-written updates on cash flow and financial health.

- **Predictive Analytics**: AI forecasts trends by analyzing past data. Businesses can anticipate customer demand, detect potential churn risks, and even predict equipment failures before they happen, allowing a proactive approach rather than reactive fixes.

With analytics powered by AI, organizations move beyond static reporting to dynamic, forward-looking decision-making.

Automation and Process Optimization

AI enhances productivity by automating repetitive tasks, optimizing workflows, and improving resource allocation, allowing employees to focus on higher-value work.

- **Robotic Process Automation (RPA)**: AI-powered RPA handles complex tasks involving unstructured data. For instance, legal departments use AI to review contracts and assess compliance risks, while banks streamline customer onboarding with AI-driven identity verification.

- **Workflow Optimization**: AI analyzes IT system logs to identify inefficiencies, revealing bottlenecks and redundancies. Telecommunications companies, for example, use AI-powered process mining to reduce service activation times.

- **Predictive Resource Allocation**: AI anticipates operational needs by analyzing historical patterns. For instance, hospitals use AI to predict patient admissions so staffing levels can meet demand.

Cybersecurity

AI strengthens cybersecurity by detecting threats faster, identifying anomalies, and automating response measures to minimize risks.

- **Advanced Threat Detection**: AI continuously monitors user behavior and network activity, flagging anomalies that indicate potential security breaches. For instance, an employee accessing large volumes of sensitive data at unusual times can trigger an investigation.

- **Automated Incident Response**: AI can take immediate action when a threat is detected, such as isolating compromised systems during a ransomware attack to prevent further damage.

- **Fraud Prevention and Anomaly Detection**: AI spots deviations from standard transaction patterns that human analysts might miss. Government agencies use AI to detect unusual network activity, while financial institutions rely on AI to identify fraudulent transactions and prevent e-commerce fraud.

Compliance and Regulatory Management

AI automates compliance tasks so businesses can stay aligned with evolving regulations while reducing manual effort.

- **Real-Time Monitoring and Reporting**: AI continuously tracks transactions and activities, instantly flagging regulatory breaches. Securities firms use AI to detect trade compliance violations, while energy companies rely on it for automated environmental reporting.

- **Regulatory Adaptation**: AI keeps businesses informed of changing laws, helping multinational corporations adjust their supply chain practices in response to shifting trade regulations.

- **Risk Assessment and Supplier Compliance**: AI evaluates suppliers and partners for compliance risks before engagement. Pharmaceutical companies, for example, use AI to verify supplier adherence to quality standards, mitigating potential legal and operational issues.

AI in Key Business Functions

AI's influence extends across all business functions, upgrading operations, sales, marketing, customer service, human resources, finance, and research. Each area benefits from AI's ability to improve proficiency, reduce risk, and accelerate innovation.

Operations and Supply Chain

AI improves predictive maintenance, inventory management, and demand forecasting. By analyzing sensor data, AI anticipates equipment failures before they occur. The result? A reduction in downtime and maintenance costs, plus an expansion of a machine's lifespan. Manufacturing plants use AI to monitor assembly lines and schedule repairs during off-peak hours.

In inventory management, AI analyzes sales trends, seasonal fluctuations, and supply chain dynamics to optimize stock levels. Machine learning models automate reordering and adjust inventory in real time, keeping popular products in stock while minimizing excess inventory. Retailers use smart systems to balance supply across stores and warehouses efficiently.

AI also aids demand forecasting by processing vast datasets, including sales history, market trends, and external factors like economic conditions or weather. Businesses can more effectively adjust production schedules and resource allocation, reducing waste. For example, an apparel company may use AI to predict seasonal demand and adjust production volumes accordingly, preventing overproduction while still meeting customer needs.

Case Study: Walmart's AI-Powered Inventory Management

Walmart implemented AI-driven demand forecasting and automated replenishment through its High-Performance Analytic Appliance (HANA) system, which analyzes sales patterns, weather, and local events. Armed with that data, Walmart reduced out-of-stock incidents by 16%, improved inventory turnover by 7.5%, and significantly lowered warehousing and transportation costs.

Marketing

AI is reshaping marketing with the way it enables precise customer segmentation, tailored promotional campaigns, large-scale comparative testing, sentiment tracking, and dynamic creative content generation. By analyzing customer data, AI identifies patterns in purchasing behavior and demographics, allowing businesses to tailor marketing strategies to specific audiences. For example, an e-commerce platform can segment customers into "bargain hunters" or "premium shoppers," delivering targeted promotions that drive higher engagement and conversions.

Personalized marketing is another key benefit of AI. AI-powered algorithms monitor customer interactions across various channels, providing highly tailored recommendations, such as streaming services using AI to recommend shows based on users' viewing histories to increase engagement and subscription retention.

AI also dynamically creates and assesses multiple versions of marketing content for different audience segments. For instance, an online retailer can experiment with various email subject lines and promotional offers, automatically choosing the most effective option to maximize engagement. This guarantees that personalized marketing efforts remain relevant.

Sentiment analysis adds another layer of insight by monitoring social media posts and customer reviews to gauge public perception of a brand. Businesses can use these insights to address concerns, refine messaging, and strengthen brand reputation. As consumer expectations evolve, companies will place even greater emphasis on brand-building and engagement strategies powered by AI.

Case Study: Coca-Cola's Marketing Shift

Coca-Cola uses AI to enhance its marketing campaigns through partnerships with NVIDIA, an AI provider, and WPP, a large advertising agency. Using the NVIDIA Omniverse platform, the company produces real-time 3D simulations and adaptive marketing content. These simulations include highly customizable product visualizations and culturally tailored meal and beverage scenarios, enhancing consumer engagement and brand affinity. Coca-Cola also employs digital twins within its Prod X studio. It enables rapid virtual testing of marketing strategies and product placements without physical prototypes, significantly reducing costs and accelerating campaign deployment. AI-driven 3D advertising assets allow swift adjustments to local market preferences, ensuring dynamic, culturally relevant content. This approach streamlines content creation, reduces production times, and maintains global brand consistency.

Sales

AI is revolutionizing sales through lead generation, sales forecasting, and pipeline management. By integrating AI-powered tools into platforms like Salesforce, sales teams can benefit from advanced insights and automation.

AI-driven lead generation and scoring analyze vast datasets to identify and prioritize high-potential leads. These systems track online behavior, engagement levels, and purchasing intent, automatically ranking prospects based on their likelihood to convert. Sales teams can then focus on leads with the highest potential, which ultimately improves conversion rates and resource allocation.

AI also strengthens sales forecasting and pipeline management by analyzing historical trends and current market conditions. Predictive analytics help businesses anticipate future sales performance, identify risks, and optimize their strategies. For example, platforms like Aviso generate WinScore Insights, which assess deal likelihood and provides data-backed recommendations to boost close rates.

Case Study: Martal Group's Sales Evolution

Martal Group, a B2B lead generation company, faced significant prospecting challenges during the COVID-19 pandemic and turned to 6sense's Revenue AI for Sales for help. In addition to providing psychographic analysis, the AI-powered platform identified prospects and boosted marketing to these specific accounts, allowing Martal Group to focus on high-potential leads. As a result, the company experienced 762% growth, improved prospect engagement with over 20% email open rates, and contributed to more than $10 million in partner revenue.

Customer Service

In the customer service field, AI-powered virtual assistants, sentiment analysis, and multilingual support are reshaping how businesses interact with customers. What that means: reduced costs and consistent service across multiple channels and geographies.

Since AI can respond to voice and text, AI virtual assistants handle routine inquiries, provide product information, and troubleshoot common issues whether the customer chooses email, voice, or chat. Automating repetitive tasks improves response times and will escalate complex issues to humans that require critical thinking and empathy, ultimately boosting customer satisfaction.

AI also plays a crucial role in understanding how customers feel by analyzing feedback from reviews, social media, and surveys. Using AI's real-time insights, businesses can quickly address concerns and refine customer interactions. Despite these benefits, balancing automation with human touch remains a challenge. While AI improves efficiency, customers still value human empathy in sensitive interactions. One solution is using sentiment analysis to detect escalating issues and seamlessly transfer high-priority cases to specialized service teams, ensuring a superior customer experience.

Case Study: Klarna's AI Customer Service Agents

An AI-powered assistant at Klarna, a Swedish fintech company, managed 2.3 million customer inquiries in its first month. The system reduced average resolution times from 11 minutes to under 2 minutes, improving customer satisfaction while maintaining service quality at scale.

Finance and Accounting

AI is revolutionizing finance and accounting in several ways. For one, AI-driven automation streamlines accounts payable processes by handling invoice matching, cross-verifying data, and flagging discrepancies. This reduces manual errors and speeds up approvals, enabling finance teams to concentrate on higher-value tasks rather than routine processing. In the past, CEOs often had to spend millions of dollars and wait months just to implement software capable of performing essential financial and business analysis. As a result, valuable data was frequently left untouched because it was too complex or costly to utilize. But today, AI simplifies these same challenges by allowing companies to create their own software solutions using everyday language instead of complicated code. Compared to traditional methods, AI-driven tools are much more affordable and quicker to deploy, giving businesses faster access to insights they can actually use.

Fraud detection is another area where AI excels. By continuously analyzing transaction patterns, AI can identify anomalies, such as unusual purchasing behavior or unauthorized access, and flag potential security threats. Credit card companies, for example, use AI to monitor billions of transactions, thereby reducing fraud and strengthening customer trust.

AI also enhances financial forecasting and risk assessment by analyzing both internal financial metrics and external economic trends. Investment firms use AI to simulate market scenarios, predict portfolio risks, and provide strategic recommendations.

Ultimately, AI is shifting finance teams from reactive to proactive decision-making, moving accounting from a back-office function into a key driver of business strategy and improvement.

Case Study: QuickLoan Financial's AI-Enhanced Loan Approval Process

QuickLoan Financial struggled with slow, error-prone loan processing that frustrated customers and overwhelmed staff. To address this, they implemented a loan approval system that used machine learning and natural language processing to analyze applicant data, assess creditworthiness, and predict default risks. The AI model included a feedback loop for continuous learning and an explainability feature to ensure transparency and compliance. As a result, loan processing times dropped by 40%, high-risk applications were detected 25% more accurately, and customer satisfaction improved with faster approvals—all while maintaining a low default rate.

Human Resources

Human resource teams are increasingly turning to AI-driven recruitment to simplify repetitive tasks like screening resumes and scheduling interviews. This frees HR professionals to focus on connecting with candidates and identifying the best talent more effectively. AI can quickly analyze thousands of applications, ensuring better alignment between job requirements and applicants while also helping reduce unconscious human bias.

AI is also transforming employee engagement by analyzing performance metrics, employee feedback, and indicators of potential turnover. By catching signs of disengagement early, HR teams can proactively introduce initiatives aimed at improving workplace culture and enhancing job satisfaction, leading to higher employee retention.

When it comes to workforce planning, AI plays a valuable role in forecasting staffing needs based on factors like business growth, seasonal fluctuations, and employee attrition. For example, retail companies rely on AI to accurately predict holiday staffing requirements, helping them maintain the right number of employees without overspending or understaffing.

Case Study: T-Mobile Builds DEI with AI

T-Mobile aimed to enhance diversity, equity, and inclusion (DEI) by creating a recruiting process that was more welcoming and accessible. To achieve this, the company adopted Textio's AI-powered language solution so job postings, recruiting emails, and branding materials conveyed a sense of belonging. By using Textio, recruiters could effortlessly implement inclusive language recommendations. The benefits were clear: Female applicants increased by 17%, job positions were filled five days faster, and recruiters gained greater DEI awareness. T-Mobile's success underscores how AI-driven language refinement can foster a more inclusive hiring process and bolster workplace diversity.

Research and Innovation

AI streamlines data analysis, simulations, and modeling, greatly speeding up product development processes. Researchers can quickly test hypotheses, validate concepts, and refine designs, resulting in faster innovation cycles and reduced time to market.

AI-powered tools analyze vast datasets, extract key insights, and run simulations without needing physical prototypes. Natural language processing (NLP) accelerates literature reviews, while predictive analytics identify trends and opportunities for new product development.

In industries such as pharmaceuticals, AI is revamping drug discovery because it can analyze biological data, predict compound interactions, and identify potential treatments at unprecedented speeds. Similarly, technology companies use behavioral analysis to upgrade product features and forecast market demands.

AI also uncovers hidden patterns that drive breakthrough discoveries. Thanks to AI-provided forecasting trends and strategic recommendations, businesses can anticipate customer needs and maintain a competitive edge.

Case Study: Pharmaceutical R&D at AstraZeneca

AstraZeneca, a global leader in pharmaceuticals, uses artificial intelligence to transform its drug discovery process significantly, making it faster and more cost-effective than traditional methods. By applying an advanced AI platform, the company quickly analyzes large volumes of data to pinpoint promising drug candidates and predict their potential effectiveness, speeding up critical phases like hit identification and lead optimization.

Using sophisticated machine learning algorithms, AstraZeneca can efficiently screen millions of compounds virtually, identifying the most promising ones for further laboratory testing. This innovative approach has notably improved success rates in early clinical trials, with Phase 1 trials for drugs discovered using AI achieving success rates between 80% and 90%, well above the industry average of 40-65%.

Additionally, the AI system analyzes biomedical literature and chemical data to fine-tune molecular structures and reliably predict drug-target interactions, streamlining the overall discovery pipeline. As a result, AstraZeneca has successfully shortened the drug discovery phase from the traditional 3-6 years to approximately 1-2 years, significantly lowering related research and development costs.

AI is the foundation of the next phase of business evolution. In the next chapter, we'll explore how AI modernizes individual business functions and what organizations must do to harness AI effectively.

Takeaways and Tips

1. **Successful AI implementation requires a clear strategy.** AI is not a plug-and-play solution; it requires thoughtful integration. Businesses must ensure they have high-quality data, clear objectives, and strong governance in place to maximize AI's potential. A structured approach to AI adoption will lead to stronger and more sustainable results.

2. **AI is affecting nearly every industry.** Through tailored AI applications, banking, retail, pharmaceuticals, healthcare, manufacturing, and technology are experiencing measurable revenue growth and performance gains.

3. **AI is reshaping business at every level.** AI is fundamentally changing how businesses operate, compete, and innovate. From streamlining workflows to enhancing decision-making, AI's influence extends across all business functions, making it a critical tool for long-term growth.

4. **Adoption speed determines competitive advantage.** Companies integrating AI today are already seeing measurable benefits, with an average return of $10.30 for every dollar spent. Organizations that hesitate will risk falling behind as AI-driven businesses become faster, more efficient, and more innovative.

Actionable Steps to Integrate AI in Your Business

- **Learn from successful AI adoption across industries**. Review AI case studies and success stories regularly within your industry and beyond to uncover practical insights and inspire innovative applications.

- **Continuously evaluate emerging AI technologies**. Monitor new AI advancements and evaluate their potential effects on customer experience and internal optimization.

- **Foster proactive curiosity**. Encourage an ongoing organizational mindset of curiosity and experimentation to quickly incorporate new AI technologies as they emerge.

SECTION TWO

Your Roadmap to AI Adoption: Process, People and Pilot Projects

CHAPTER 5:

Is Your Organization Ready for AI?

The adoption of AI requires a strong foundation of strategy, infrastructure, and culture. So, how do you know if your organization is truly prepared?

In this chapter, you'll learn:

- How to assess your organization's AI readiness across strategy, processes, talent, and culture.

- Why well-documented workflows are essential for AI implementation.

- The role of leadership in fostering AI adoption and managing change.

- How governance, infrastructure, and pilot programs set the stage for AI success.

A strong AI strategy starts with a clear vision, supported by robust operational processes and a workforce prepared to embrace change. Without these elements, even the most advanced AI initiatives risk failing. This chapter will guide you through a comprehensive AI readiness assessment, helping you identify strengths, address gaps, and build a solid foundation for effectively integrating AI into your organization.

Strategic Vision and Growth Plan

The foundation of any AI strategy is a clear and ambitious vision for the future. I harnessed the Entrepreneurial Operating System (EOS) and its core principles at my previous company to establish a strategic vision and confirm that the teams were aligned. The core focus of EOS helped us determine our company's passion, niche, and strengths, helping us to stay focused on essential activities. We also set clear long-term goals, such as the 10-Year Target and the 3-Year Picture, which jibed with our vision. We used short-term objectives like the 1-Year Plan and Quarterly Rocks to make the vision actionable and to drive accountability and progress.

Focusing on core values, goals, and purpose helps you to deploy AI in a way that meets your organizational objectives and spurs meaningful progress. Many of the basic principles of EOS, such as documenting strategic goals and maintaining focus on key priorities, fold seamlessly into preparing for AI implementation. Start by documenting your strategic goals: Where do you want to be in five or ten years? The next step is to match AI initiatives directly with these goals so they contribute to objectives like market expansion, improved customer experience, or operational output.

Documented Processes from Sales to Operations

For AI to make a meaningful impact, you need a strong operational foundation. This starts with having well-documented processes, particularly in key areas like sales and operations. Process mapping, standard operating procedures (SOPs), and data flow documentation are all critical. A few key questions to examine:

- Have you documented the 20% of processes that produce 80% of your business's results? Most companies have 6 to 10 core business processes that are essential to their operations. These typically include human resources, marketing, sales, operations, finance, and customer support.

- Do all processes have an owner who is responsible for getting feedback to create clear, repeatable steps that ensure consistency?

- Are the processes centralized, making them easily accessible to all team members? Does everyone have access to the most up-to-date information?

- Is everyone in your organization following the documented processes?

Clear, consistent processes create a fertile ground for AI solutions to thrive, leading to operational excellence and happier customers. However, even the most sophisticated AI systems require the right people to guide their implementation and maximize their potential. Thus, you need to cultivate the necessary talent and skills so your workforce can collaborate effectively with AI and create meaningful change.

Talent and Skills

AI is more than just algorithms; it's about people. The success of your AI projects heavily relies on the talent within your organization. Assessing your current workforce's skills in areas like data analytics, machine learning, and AI is critical. It's also important to emphasize general technical skills that allow employees to adapt to new technologies and processes. You should evaluate your workforce's proficiency in data literacy, problem-solving, and critical thinking. Fostering a mindset of continuous learning and adaptability will prepare your team for AI tools and applications.

Once you understand your talent landscape, you can bridge gaps through targeted training and recruitment. This could involve upskilling current employees, forming partnerships for external training, or hiring new talent with specialized skills. Recruitment strategies should focus on attracting individuals who not only possess technical expertise but also have strong communication skills and the ability to work across functions.

Organizational Culture

The best technology won't succeed if your culture opposes change. AI adoption requires a culture that values experimentation while actively involving employees in the process. Fostering such a culture means encouraging team members to contribute ideas, listening to concerns, and being transparent about the benefits and challenges AI may bring.

Success also hinges on the technology and data infrastructure that supports these initiatives. Robust systems and well-structured data are the backbone of effective AI implementation, ensuring that cultural readiness is matched by technical capability.

Technology and Data Infrastructure

When it comes to AI, having the right tools in place is like having a solid foundation for a house. Your hardware, software, and data systems must work together seamlessly to support your AI initiatives. Think of scalability as your safety net: Your infrastructure needs to keep pace as your AI projects expand. That means handling larger datasets, supporting more advanced models, and staying flexible for what's next. With the right setup, you'll be ready to tackle growing demands and unlock AI's full potential.

Readiness Assessment

The next step is to examine your organization's overall readiness for AI adoption. The following assessment is designed to help you evaluate critical areas, from strategic vision to governance and pilot programs, so you can be confident that you're equipped to maximize AI's potential and get the results you want.

In other words, think of this assessment as a road map checkpoint, a way to gauge how prepared you are to integrate AI into your operations effectively. By scoring each category, you'll uncover your strengths and the areas that need attention, giving you a clear path forward for seamless and successful AI adoption. Share the question-naire with your leadership team and ask them to complete the assessment as well.

Rate each question on a scale of **1 to 5**, where:

1 = Not at all

2 = Needs significant improvement

3 = Somewhat

4 = Mostly

5 = Fully implemented

Category	Question	Rating (1-5)
Strategic Vision and Growth Plan	Does your organization have a clearly defined strategic vision?	
	Have you documented long-term (5-10 years) and short-term (1-year, quarterly) goals?	
	Have you identified how AI initiatives may help you meet your business goals?	
	Have you considered how AI initiatives align with core business objectives such as market expansion or operational efficiency?	
Documented Processes	Have you documented the key business processes (e.g., HR, Marketing, Sales, Operations, Finance, Customer Support)?	
	Are there designated owners for each core process responsible for feedback and optimization?	
	Are all documented processes easily accessible and centralized for all team members?	
	Do all employees consistently follow the documented processes?	
Talent and Skills	Does your workforce have proficiency in leveraging data to make decisions?	
	Are employees proficient in data literacy, problem-solving, and critical thinking?	

Talent and Skills	Is there a culture of continuous learning and adaptability within your organization?
	Do you have training and development programs that could be used for AI-focused training?
	Are your recruitment strategies aimed at attracting individuals with technical acumen and strong communication skills?
Company Culture	Does your organization value innovation and encourage experimentation with new technologies?
	Are employees actively involved in transformation efforts?
	Is there transparent communication about the benefits and challenges of new initiatives?
	Do employees trust leadership and show buy-in toward new initiatives?
Data Governance and Leadership	Do you have strong support from senior leadership for adopting AI in your organization?
	Have you established an AI governance framework?
	Do you have the infrastructure to store, process, and analyze large volumes of data?
	Is your data accurate, up-to-date, and easily accessible for AI models?
	Have you conducted a risk assessment for AI implementation?

Data Governance and Leadership	Do you have strategies to mitigate identified risks and ensure compliance with regulations?	
	Have you analyzed your existing agreements with your customers to determine whether there are any issues with using AI (e.g., data ownership, copyright concerns around work produced with AI)?	
Technology and Tools	Have you acquired the necessary technology stack for AI?	
	Have you considered how your systems integrate with AI solutions?	
	Do you have measures to identify and mitigate biases in AI models?	
Operational Readiness	Do you have budget for AI initiatives?	
	Do you have ROI metrics that can allow a benchmark for AI projects?	
	Do you have a change management plan to address potential resistance?	
Total		_____/150

Assessment Interpretation:

120-150: Highly Ready. Your organization is well-prepared for AI implementation. Focus on fine-tuning strategies and scaling AI initiatives.

90-119: Moderately Ready. Your company has a solid foundation but needs to address specific areas for successful AI adoption.

60-89: Low Readiness. Significant improvements are needed in multiple areas before your organization should pursue AI implementation.

30-59: Minimal AI Readiness. Foundational work is needed.

There's no need to worry if your assessment score indicates that your organization isn't yet ready to implement AI. In the upcoming sections of this book, you'll find a wealth of resources and actionable tips designed to guide your organization toward AI readiness and a smooth and successful integration.

AI adoption is not a one-time initiative; it is an ongoing evolution that requires continuous adaptation. As organizations incorporate AI into their operations, leadership plays a crucial role in steering this transition.

In the next chapter, we will delve into the changing demands of AI leadership and analyze how business leaders can manage this shift in a manner that delivers long-term value while upholding trust and accountability.

Takeaways and Tips

1. **A strong foundation is essential for AI success**. AI is not a one-size-fits-all solution. It requires alignment with strategic goals, a robust infrastructure, and a culture that supports curiosity. Organizations that assess their readiness and address gaps in processes, talent, and governance will be best positioned for successful AI adoption.

2. **AI readiness extends beyond technology**. Businesses must document key processes, develop a culture of adaptability, and equip employees with the skills to collaborate effectively with AI. Without these elements, even the most advanced AI systems will struggle to deliver breakthroughs.

3. **Organizational culture determines the speed of AI adoption**. Companies that foster a culture of experimentation and continuous learning will integrate AI more effectively. Leadership buy-in, transparent communication, and employee engagement are critical for overcoming resistance and driving growth.

4. **Governance and data quality are non-negotiable**. AI can only be as effective as the data it relies on. Clear governance structures, ethical guidelines, and well-maintained data infrastructure are crucial for ensuring AI operates responsibly and delivers reliable insights.

Actionable Steps to Prepare for AI

- **Conduct the AI readiness assessment in this chapter.** Evaluate your organization's strategic vision, documented processes, talent capabilities, and technology infrastructure to identify strengths and areas for improvement.

- **Strengthen process documentation.** Clearly map out essential workflows to create a structured foundation for AI integration. Make sure key processes have owners who can oversee AI implementation.

- **Invest in AI-related skill development.** Upskill employees in data literacy, problem-solving, and AI-driven decision-making to enhance adoption and minimize resistance.

- **Establish strong governance frameworks.** Define ethical guidelines, data management policies, and AI oversight structures to secure responsible implementation.

- **Start with targeted pilot programs.** Test AI applications in specific areas before scaling across the organization, allowing for controlled experimentation and refinement.

CHAPTER 6:
Leading AI Transformation

Guiding an organization's AI modernization requires a new kind of leadership. This leadership must embrace AI's potential while ensuring its responsible and ethical deployment.

In this chapter, you'll learn:

- How AI leadership differs from past technological shifts.

- The role of AI in shaping business strategy and long-term vision.

- Why a clear AI philosophy is essential for guiding decision-making.

- The importance of stakeholder buy-in of AI.

AI's rapid evolution means business leaders must think beyond short-term gains and develop a long-term approach to AI. Establishing an AI philosophy, involving stakeholders, and implementing governance structures are critical to using AI responsibly and effectively.

This chapter will guide you through the foundational elements of AI leadership and help you understand how to create a governance framework that balances progress with ethical considerations.

A Bigger Technology Evolution Than Before

Every executive today has experienced past technological breakthroughs, such as the rise of personal computing, the internet boom, mobile technologies, and cloud computing, among others. However, AI represents a revolution of an entirely different magnitude. It's not hyperbole to say that AI's influence will likely dwarf that of any previous tech shift. Why? Because AI has the potential to transform nearly every aspect of how businesses operate and how consumers interact with the world. Over recent decades, we have digitized and collected vast amounts of information but have lacked the tools to fully analyze or leverage this data effectively. AI addresses this gap, shifting us from an era of process automation (machines following predefined rules) to one of intelligent automation and augmentation (machines that can learn, reason, and make decisions). The scale and scope of this change have never been seen before.

A landmark study by the auditing firm PwC projects that AI could contribute up to $15.7 trillion to the global economy by 2030, making it one of the greatest drivers of growth in history, surpassing the current output of China and India combined. Similarly, Stanford's AI experts predict an impending productivity boom, with AI sparking a business revolution "more disruptive than any other tech-driven transformation in history." In simple terms, AI is a general-purpose technology (like electricity or the internet) that has fundamentally reshaped the business landscape. Just as electrification in the early 20th century completely changed industry, AI is set to become the invisible backbone of innovation and operations in the 21st century.

Consider the scale of change in three areas: consumer experience, operations, and previously impossible capabilities:

- **Customer Experience**: AI allows an extreme level of personalization and responsiveness that was never possible before. Companies can deliver Netflix-style recommendation engines for everything, tailoring products, services, and content to each individual's tastes in real time. Customer service bots handle millions of inquiries 24/7, giving instant answers and support. In retail, AI can predict what customers want even before they know it, enabling proactive offerings. We're also seeing the rise of entirely new experiences, such as virtual assistants that can plan our schedules and augmented reality shopping. The result is a leap in customer expectations; people will gravitate toward businesses that offer the speed, personalization, and convenience that only AI can provide.

- **Operations and Decision-Making**: Internally, AI is turbocharging how companies run. Routine office processes (invoice processing, scheduling, data entry) can be automated with AI, dramatically reducing costs and error rates. More impressively, AI systems can analyze vast datasets to find patterns and optimize operations in ways humans would never spot. Supply chains can be dynamically rerouted in response to real-time conditions; predictive algorithms can adjust manufacturing output in anticipation of demand fluctuations; maintenance can shift from scheduled checks to AI-predicted just-in-time repairs. Even strategic decision-making is being transformed; executives now have AI-powered analytics to evaluate scenarios or investment options with unprecedented depth. It's as if every department has gained a super-intelligent co-pilot. The net effect is faster, smarter operations at scale; enterprises can respond instantaneously to changes and continually self-optimize.

- **Previously Impossible Capabilities**: AI is enabling businesses to achieve things that were pure science fiction just a decade ago. We now have AI models capable of designing new chemical compounds, potentially leading to drugs for diseases that previously had no cure. Machines can understand and generate human-like language, facilitating automated report writing and marketing content creation. In technology development, AI can even write code and assist engineers in solving complex design problems.

  Tasks once considered inherently human—such as driving a car, composing coherent articles, or recognizing faces and speech—are now being performed (and sometimes surpassed) by AI systems. This opens the door to entirely new products and services. For instance, researchers at the Yale School of Medicine developed a new AI algorithm that can accurately detect two types of deadly cardiomyopathies, on average, two years before patients receive formal diagnoses. This new AI tool utilizes fast, inexpensive ultrasounds already collected during ER visits, allowing for earlier detection without placing an additional burden on healthcare providers.

Executives must grasp that AI is not a singular invention but a constellation of advancing technologies (machine learning, natural language processing, computer vision, robotics, etc.) all maturing around the same time. This compounding effect means AI's impact is broad and accelerating. Consumer adoption of AI-powered services is also lightning-fast; consider how a generative AI tool like ChatGPT gained over 100 million users in just a couple of months after launch, reflecting a ravenous appetite for these kinds of solutions.

The takeaway: AI will change virtually every aspect of business. The companies that thrive will be those whose leaders recognize this sweeping scope and move decisively to harness AI across the organization.

Your Organization's Reimagined Vision

Take some time to imagine what an "AI-transformed" version of your company might look like in 5 or 10 years. How would products, services, and operations be different? Your vision might be to have a fully data-driven supply chain that self-optimizes instantaneously, or to offer a hyper-personalized customer experience. Having this North Star vision is important because it ensures all the small steps you take work toward a meaningful goal. Communicate this vision internally so everyone knows why you're investing in AI.

A few questions to consider:

1. How does what you offer change when we have AI that is more advanced than PhDs in their own field?

2. How does your vision shift as we achieve artificial super intelligence or AI that is more capable than any human?

3. How does my model shift if I am selling indirectly through an AI agent?

4. What are the emerging needs of this technology that your company may be able to serve?

5. What are some of the challenges faced by your business or customers that you haven't been able to solve?

Thinking Big but Starting Small

Given AI's expansive potential, executives must think big to develop a bold vision for how AI could reinvent their company, but they also need to start small to achieve that vision effectively. Trying to implement a sweeping, organization-wide AI overhaul in one go is a recipe for disappointment. The wiser approach is to focus initially on projects that offer clear rewards with minimal risks to reputation or critical operations. These low-risk, high-impact projects provide opportunities to gain competency, build internal capabilities, and learn valuable lessons. Recognizing that your organization must crawl before it can run enables each incremental win to strengthen the foundation for increasingly ambitious AI initiatives. Over time, this approach transforms early successes into a robust framework capable of supporting comprehensive, transformative change.

Defining Your Company's AI Philosophy

Given the speed at which AI is evolving, a solid and clear AI philosophy creates a safety net for adapting to change.

These guiding principles act as a compass, fostering trust among all stakeholders: employees, customers, partners, and the broader community.

Alison Lami, a recognized AI thought leader, puts it succinctly: "You can play possum and pretend the world isn't changing, or you can adapt: either incorporate AI into your existing business or create a new AI-native model." Her words emphasize the necessity for a clear, proactive approach to AI that matches your organization's values and strategic direction.

Key Principles of an AI Philosophy

1. **Transparency**: Confirm employees and customers understand how AI is being used. Clear policies foster trust and allow people to make informed decisions.

2. **Accountability**: Establish who manages AI, including identifying and mitigating risks. If something goes wrong, there's a clear way to address it.

3. **Human-Centered Design**: Use AI to enhance human capabilities, not replace them. Always consider the social and ethical consequences of AI.

Creating Your AI Philosophy

The process of defining an AI philosophy involves:

- **Identifying core values:** Fit your AI use with your company's mission, such as inclusivity, or privacy.

- **Establishing ethical guidelines:** Define acceptable uses of AI and have a plan for addressing issues like bias or privacy concerns.

- **Ensuring transparency and accountability**: Make sure AI decisions are explainable and outline who monitors AI's impact.

- **Embracing human-centered design**: Seek AI solutions that genuinely benefit people and enhance their experiences.

Real-World Examples

- **Google**: Their AI principles focus on responsible development and transparency, making sure AI is aligned with human values.

- **Microsoft**: They emphasize fairness, inclusiveness, and accountability and have an AI Ethics Committee to oversee initiatives.

- **IBM**: They prioritize transparency, fairness, and data privacy, all critical aspects, especially in sensitive industries like healthcare.

Why It Matters

The groundwork you lay today will set the direction for your AI initiatives. Whether it's establishing a robust governance structure or understanding AI's impact on employees, customers, partners, and society, the AI Governance Council—the core group typically composed of an AI Leader, representatives from Legal and Compliance, and an Executive Sponsor—acts as the nerve center of these efforts. This council bridges the gap between technical advancement and ethical considerations, fostering cross-departmental collaboration. We'll dive deeper into its structure and role in the next chapter.

However, even the most thoughtful AI strategies require executive and financial support to thrive. AI leaders must anticipate and address key concerns, ranging from investment costs to measurable results, to secure buy-in and drive adoption initiatives forward.

Aligning Strategy, Investment, and Execution

While many executives recognize AI's potential, they often struggle with justifying the investment and seeing how AI initiatives match with long-term strategic goals. Successful AI leaders focus on three key areas when securing buy-in:

- **Clear Business Value**: Demonstrating AI's significance beyond automation, such as new revenue models, competitive differentiation, and market expansion.

- **Strategic Alignment**: Ensuring AI initiatives support broader business goals rather than being siloed experimental projects.

- **Financial Feasibility**: Addressing cost concerns and leveraging available funding incentives to offset AI investment risks.

Framing AI as a Business Imperative, Not Just a Technology Investment

One of the biggest challenges in securing AI buy-in is shifting the conversation from technology-first thinking to business-first impact. Leaders who successfully implement AI do not position it as an IT upgrade; they frame it as a fundamental shift in how the company creates value, improves decision-making, and builds resilience in a rapidly evolving market.

For AI initiatives to gain executive and stakeholder support, they must be positioned as a strategic necessity, not an optional experiment. A well-structured AI strategy answers the following key questions:

- How does AI directly contribute to business growth, customer experience, or operational excellence?

- What is the cost of inaction? How will competitors who embrace AI gain an advantage?

- How does AI fit within our broader strategy?

By connecting AI to tangible business outcomes such as revenue growth, customer retention, and competitive agility, leaders can strengthen the case for investing in AI.

Don't Overlook Financial Incentives: R&D Tax Credits and AI Investment Support

AI-driven innovation requires investment, but many organizations overlook financial incentives that can help offset costs. One often-underutilized opportunity is Research and Development (R&D) tax credits, which governments offer to encourage technological advancement.

Many AI initiatives qualify for R&D tax credits, including:

- Developing custom AI solutions tailored to business needs.

- Piloting experimental AI models or integrating new AI tools.

- Enhancing existing processes with innovative approaches.

By adding R&D tax credits into the AI investment strategy, organizations can recover a portion of development costs, including software, employee training, and AI experimentation. This financial cushion reduces upfront risk.

Beyond R&D credits, businesses should also explore:

- AI grants available through government or industry programs.

- Public-private partnerships to co-develop solutions.

- Tax incentives for AI infrastructure investments, such as cloud computing and data analytics tools.

With a strong leadership foundation in place, the next step is to move from strategy to execution.

The following chapters will explore how businesses can use AI to deliver tangible results while staying true to the company's broader vision. We'll start by exploring how to identify and develop AI leaders—your "AI unicorns"— who possess the unique combination of skills needed to drive AI implementation successfully. By factoring these financial levers into AI strategy discussions, executives can address investment concerns while demonstrating a commitment to fiscal responsibility.

Takeaways and Tips

1. **AI leadership extends beyond technology**. AI is a business-wide transformation. Effective AI leaders must bridge technical expertise with strategic vision so AI initiatives drive real business value while maintaining ethical and responsible implementation.

2. **Strategic vision is essential for AI success**. AI is evolving rapidly, and leaders must think beyond short-term gains. The best AI leaders develop a long-term vision, aligning AI initiatives with the company's broader business strategy to achieve sustainable growth and competitive advantage.

3. **Stakeholder alignment is key to AI adoption**. AI success requires buy-in from leadership, employees, and external stakeholders. Leaders must proactively address concerns, communicate AI's benefits, and foster a culture of trust and collaboration.

4. **An AI philosophy creates stability amid rapid change**. With AI advancing at an unprecedented pace, a strong AI philosophy provides a framework for consistent decision-making, ethical considerations, governance, and long-term adaptability, preventing reactionary or misguided AI adoption.

5. **Maximize AI value through strategic investment**. While AI adoption requires a financial commitment, organizations must coordinate their investments with long-term value creation. Leaders should explore financial incentives like R&D tax credits and grants to maximize returns while effectively managing costs.

Actionable Steps to Strengthen AI Leadership

- **Define and communicate your AI vision with leadership.** Develop a vision for your organization's stance on AI that reflects your company's mission and goals. Make sure leadership and teams understand AI's role in shaping business strategy.

- **Create your company's AI philosophy.** Adapt our template (https://www.demysti-fai.com/aimadesimple/execlibrary) to begin building your AI philosophy by clearly defining ethical guidelines, transparency, and accountability.

- **Maximize financial incentives to support AI investment.** Identify and apply for R&D tax credits, AI grants, and industry funding to offset costs and accelerate adoption. Make sure your financial stakeholders understand AI's return on investment beyond immediate cost savings.

CHAPTER 7:
Finding Your AI Unicorn

When you're ready to move your organization's AI strategy from concept to implementation, your first step is to appoint a leader for these initiatives. This person is typically someone in a role such as Chief AI Officer, Chief Innovation Officer, or Chief Digital Officer. Regardless of their title, you want a strategic thinker who not only understands your business and industry but also is comfortable with technology and adept at inspiring and winning people over. To be honest, you are looking for a unicorn.

In this chapter, you'll learn:

- The key traits that define an AI leader, or "AI unicorn."

- Why curiosity, strategic thinking, and communication skills matter as much as technical expertise.

- The responsibilities of an AI leader in driving AI adoption, governance, and innovation.

- How to decide between hiring a full-time AI leader or leveraging a fractional AI executive.

- The benefits of hiring AI leaders from diverse backgrounds beyond traditional technical roles.

- How to rethink AI job descriptions to attract the right talent.

While the ideal AI leader has machine learning expertise or a background in data science, they also need to be a bridge between AI's technical complexities and the business's strategic objectives. They should be able to explain AI concepts to executives, drive cross-functional collaboration, and anticipate the ethical and operational implications of AI adoption.

By rethinking traditional hiring practices and broadening the talent pool beyond strictly technical roles, organizations can discover and cultivate leaders who offer fresh perspectives on AI. Whether you are hiring a Chief AI Officer, an AI strategist, or a fractional leader, this chapter will assist you in finding the right individual to lead your organization's AI journey.

Skills of a Unicorn

This "AI unicorn" is critical for spearheading advancements and integrating AI technologies into an organization. Below are the essential characteristics that define an AI unicorn:

1. **Insatiable Curiosity**: AI leadership requires a relentless drive to explore new ideas and technologies. Ideal candidates must be lifelong learners who continuously seek knowledge and stay current with developments in AI and related fields. This curiosity propels ingenuity, allowing the organization to stay ahead of emerging trends and leverage AI's potential.

   Example: Michelle Obama (Former First Lady of the United States): Though not in tech, her curiosity-driven initiatives, such as Let's Move! and Reach Higher, show her ability to tackle new challenges, connect diverse ideas, and build dynamic programs. This mindset is transferable to AI leadership, where curiosity often leads to pioneering initiatives.

2. **Analytical and Strategic Mindset**: AI unicorns must excel at breaking down complex data sets and finding actionable insights. They should be able to connect these insights with broader business strategies so AI initiatives support organizational goals and deliver measurable value. This strategic thinking allows the AI leader to prioritize high-impact projects.

   Example: Theo Epstein (Former President of Baseball Operations, Chicago Cubs): With no AI background, Epstein revolutionized team management by applying advanced analytics to build championship-winning teams. His ability to synthesize data insights into actionable strategies demonstrates the type of strategic thinking AI leaders need.

3. **Fearless Innovation**: These leaders are comfortable with uncertainty and willing to take calculated risks to ignite breakthroughs. They foster a culture of experimentation within their organizations and lead with confidence, even when navigating challenges.

   Example: Reed Hastings (Co-Founder of Netflix): Hastings innovated by introducing the process of mailing DVDs to customers, reinventing the movie rental industry. As technology evolved, he embraced change once again by shifting to online streaming, eventually discarding the physical mailing process and the iconic red envelopes Netflix once relied upon. Hastings revolutionized the entertainment industry further through AI-powered content recommendations and sophisticated streaming algorithms. His willingness to embrace risk and challenge traditional models remains a hallmark of industry disruptors.

4. **Influential Communication**: AI leaders must clearly articulate complex AI concepts across all levels of the organization, from technical teams to executives, fostering collaboration and securing buy-in for AI initiatives.

   Example: Jacinda Ardern (Former Prime Minister of New Zealand): Ardern's transparent and empathetic communication style during crises highlights how influential communication can inspire trust and unity, essential traits for leading change.

5. **Technical Proficiency (Coding Experience a Plus)**: While not always mandatory, coding experience helps AI leaders understand the technical aspects of AI and communicate more effectively with their teams. With a solid foundation in machine learning, data analytics, and automation, they can make informed decisions about technology adoption.

   Example: Satya Nadella (CEO of Microsoft): Nadella's technical background in computer science has been instrumental in guiding Microsoft's evolution, from Azure AI to AI-powered Office tools.

The examples above span industries from tech to fashion, sports, and politics, emphasizing that successful AI leaders are not confined to traditional AI/ML backgrounds. What sets these leaders apart is their ability to combine curiosity, strategic vision, ideation, communication, and (where applicable) technical fluency to unlock AI's progressive potential. These traits, rather than their specific career paths, define what it means to be an "AI unicorn" and underscore the universality of effective leadership in the modern world.

Responsibilities of an AI Unicorn

AI leaders are responsible for guiding the strategic integration of AI into an organization. Their role extends across several domains, from policy development to team management. Key responsibilities include:

1. **Creating AI Policies and Roadmaps**: AI leaders must develop comprehensive policies that address ethical concerns, such as data privacy, transparency, and accountability. They should also establish an AI roadmap that outlines key initiatives, timelines, and milestones that meet the organization's long-term goals.

2. **Drafting Communications for RFPs and Contracts**: Clear and concise communication materials are critical when interacting with external stakeholders. AI leaders are responsible for articulating AI's capabilities and benefits in requests for proposals (RFPs) and contracts so that client expectations are met.

3. **Leading Cross-Departmental Task Forces and AI Governance**: AI integration often spans multiple departments. AI leaders must establish and lead cross-departmental task forces for successful AI adoption and compliance with established governance policies. They should also implement governance structures to monitor AI systems regularly.

4. **Outlining AI Pilot Programs and Evaluating Software**: AI leaders must identify pilot programs where AI can make the most significant change. They should oversee the testing and validation of AI solutions and evaluate software for its suitability to the organization's needs, making sure that the right tools are selected for each initiative.

5. **Keeping the Organization Updated on AI Trends**: AI leaders must regularly update the team on developments, sharing insights from industry publications, conferences, and professional networks.

6. **Leading AI Product Development**: AI leaders work closely with product development teams to integrate AI into the organization's offerings so AI solutions enhance both products and service delivery.

Finding the right AI leader is only the first step. Equally important is deciding how to integrate their expertise into your organization's structure. Should you bring someone on full-time to lead the charge or consider a fractional leader who can provide specialized guidance while allowing flexibility? The choice between full-time and fractional leadership depends on your organization's current needs, resources, and long-term vision. Let's explore how to determine the best fit for your AI journey.

Full-Time vs. Fractional AI Leadership

The decision on whether to hire full-time or fractional AI leaders largely depends on factors such as the scale of AI ambitions, organizational size, and available resources. Here are some key considerations to help organizations make the right choice.

Criteria	Full-Time AI Leadership	Fractional AI Leadership
Business Size	Best for large enterprises (250+ employees); can manage complex, multi-departmental AI integrations.	Ideal for small to medium-sized businesses (SMBs); suitable for early-stage AI adoption or limited resources.
AI Ambitions and Complexity	Necessary for high AI ambitions; supports developing proprietary AI and implementing complex systems.	Works well for moderate AI needs; focuses on specific use cases, pilot projects, or incremental adoption.
Budget and Resources	Requires a substantial budget; provides dedicated, hands-on leadership for comprehensive, long-term strategies.	More cost-effective for limited budgets; offers specialized expertise on a part-time basis.
Decision-Making Framework	Suitable when AI goals are broad, complex, and long-term; requires deep integration and ongoing engagement.	Best when focusing on targeted projects or pilot initiatives; allows flexibility and scalability based on available resources.
Benefits	Fully integrated leadership dedicated to driving comprehensive AI strategies; long-term commitment to AI initiatives.	Provides expert advice without the full-time cost commitment; flexible engagement scaled to project needs.
Drawbacks	Higher overall cost; requires continuous support and full-time commitment.	Limited availability and integration; may not support broad, long-term AI strategies as effectively.

Think Differently to Hire Better

Effective AI leadership requires rethinking traditional hiring approaches. While technical expertise remains crucial, prioritizing it exclusively limits creative, strategic, and ethical thinking. Companies should consider leaders with diverse backgrounds in philosophy, ethics, or innovation who bring critical thinking skills and understand AI's human implications, creating a holistic leadership approach.

This connects to the "AI unicorn" concept: By recruiting from unconventional backgrounds, organizations can build teams that integrate technical advancements with broader organizational and societal goals.

Four Reasons to Broaden the Hiring Lens

1. **AI Is Not Just Technical, It's Philosophical**: It shapes not only business processes but also societal norms, human behavior, and ethical frameworks. Leaders with a philosophical mindset are uniquely positioned to grapple with questions about autonomy, agency, and the long-term consequences of outsourcing human judgment to algorithms. Thinkers like Oxford's Nick Bostrom and Australian philosopher Peter Singer have emphasized the ethical dimensions of AI, underscoring the need for leaders who can pair strategic insight with moral foresight to future-proof their organizations.

2. **Visionary Thinking Inspires Transformation**: Leaders from creative or philosophical backgrounds often excel at connecting disparate ideas and challenging traditional norms. Consider Steve Jobs: Though not a technical expert, his innovative mindset reshaped industries. AI leaders with similar traits can focus on

new value propositions, enabling businesses to use AI for noticeable advancements rather than incremental improvement.

3. **Understanding Business and Human Behavior**: AI's success depends on integration with human roles and customer experiences. Leaders with backgrounds in psychology, sociology, or philosophy anticipate people's responses to AI, asking how it can enhance rather than replace human roles and foster trust. They create user-centered designs that resonate with customers and employees.

4. **Ethical Governance and Long-Term Thinking**: Leaders trained in ethics excel at identifying risks and advocating for transparency, accountability, and fairness. Organizations like IBM integrate ethicists into their AI Ethics Board to address broader societal implications, embedding these principles into AI initiatives while mitigating reputational and regulatory risks.

Rethink the Job Description

The following job description reflects the need for creative, forward-thinking leaders who can navigate AI's ethical and strategic challenges and responsibly shape its role in business.

Job Title: Chief AI Strategist

Reports to: CEO/Executive Leadership

Job Summary: The Chief AI Strategist will lead the company's AI initiatives, focusing on the strategic integration of AI technologies with an emphasis on human-centered and ethical AI development. This role requires a leader who can navigate the philosophical, ethical, and social dimensions of AI, ensuring that AI not only meets technical requirements but also matches the company's broader goals and societal responsibilities. The ideal candidate will be an innovative thinker with a strong understanding of consumer behavior, business strategy, and AI's ethical implications.

Key Responsibilities:

- Develop and implement the company's AI philosophy, ensuring that AI fits with its values and long-term vision.

- Lead the AI strategy with a focus on innovation, ethical considerations, and customer-centered design.

- Ensure AI solutions address real business problems while enhancing customer and employee experiences.

- Collaborate with various departments to create a strategic roadmap that unifies AI projects with company goals.

- Oversee the development of AI policies, focusing on transparency, algorithmic fairness, and data governance.

- Advocate for responsible AI use, ensuring ethical considerations are at the forefront of all AI initiatives.

- Engage with external partners and thought leaders to keep the company at the forefront of AI developments.

Qualifications:

- Background in philosophy, ethics, social sciences, or a related field with a strong understanding of AI technologies.

- Proven track record of propelling change and leading large-scale AI initiatives.

- Ability to balance technical acumen with an understanding of human behavior and business strategy.

Key Performance Indicators (KPIs)

The success of this AI unicorn can be assessed across key strategic areas and measured through specific KPIs.

AI Strategy and Implementation

- **AI Adoption Rate**: Percentage of business units integrating AI solutions within the first 12 to 24 months.

- **Alignment with Business Goals**: Percentage of AI initiatives directly supporting key business objectives (e.g., revenue growth, cost reduction, efficiency improvements).

- **Time-to-Value for AI Projects**: Average time taken from AI project initiation to measurable impact (e.g., improved processes, increased revenue).

- **Cross-Departmental AI Collaboration**: Number of AI initiatives successfully deployed across multiple business functions.

Innovation Products and Services

- **AI Pilot Success Rate**: Percentage of AI pilot programs that transition into full-scale implementations.

- **New Product and Service Introductions**: Number of new AI-powered products, features, or service enhancements launched annually.

- **AI-Generated Revenue**: Percentage of company revenue attributed to AI-driven discoveries.

- **AI Integration in Decision-Making**: Number of strategic business decisions informed by AI insights or automation.

Ethical AI and Compliance

- **Bias and Fairness Audits Conducted**: Frequency of AI bias reviews and percentage of AI models passing fairness and compliance standards.

- **Regulatory Compliance Score**: Adherence to AI-related regulations (GDPR, CCPA, AI Act, etc.), measured by compliance audits.

- **AI Governance Implementation**: Number of governance policies established and percentage of AI projects adhering to ethical AI frameworks.

Operational Efficiency and Cost Savings

- **Cost Reductions**: Measurable reduction in operational costs due to AI automation and optimization.

- **AI Productivity Gains**: Percentage improvement in employee productivity due to AI-enhanced workflows.

- **AI Infrastructure Optimization**: Capability improvements in AI model deployment and computing resource utilization.

Stakeholder Engagement and Education

- **Executive and Employee AI Training Participation**: Percentage of leadership and employees completing AI training programs.

- **Stakeholder Satisfaction Score**: Measured through surveys assessing AI's effect on internal teams, customers, and partners.

- **AI Thought Leadership Contributions**: Number of industry events, white papers, or articles published to position the company as a leader in AI.

Building Your AI Governance Council

With your AI lead in place, the next major task is to build the structure that supports responsible AI growth. This begins with establishing an AI governance council. This dedicated group is responsible for overseeing AI projects and guiding them toward success strategically and ethically. They help set the foundation for AI success by defining clear roles and responsibilities, creating standardization and accountability across all AI efforts so that:

- AI investments meet business priorities rather than becoming unfocused pilot programs.

- AI adoption follows ethical and regulatory guidelines, avoiding reputational and compliance risks.

- Teams have clear accountability frameworks to measure AI performance and ROI.

AI governance is about making AI investments that are sustainable, scalable, and strategically guided rather than reactive or experimental. By building a governance council, you help create transparency, accountability, and consistent stakeholder trust, key ingredients to AI success.

Composition of the AI Governance Council

The AI governance council should be a diverse group of leaders who collectively understand AI from multiple perspectives. Key members might include:

- **AI Lead**: This unicorn acts as a bridge between technical execution and the broader business strategy.

- **Executive Representative**: Provides strategic oversight, ensuring AI supports the company's vision and gets the needed support without becoming entangled in day-to-day details.

- **IT Representative**: Focuses on technical feasibility and that AI tools integrate seamlessly with your current infrastructure.

- **Legal and Compliance Expert**: Monitors regulatory compliance and ethical considerations, mitigating legal risks.

- **Ethics Officer or HR Representative**: Safeguards that AI adoption is fair and supports workplace culture.

With this diverse composition, AI initiatives are examined from all necessary angles—strategic, technical, ethical, and legal—resulting in a balanced and well-governed approach. Encouraging constructive debate among council members allows potential issues to emerge early, leading to more thoughtful and resilient AI deployments.

Responsibilities of the AI Governance Council

Establishing an AI governance council should not be viewed simply as a procedural step. It's a strategic cornerstone for embedding AI successfully within your organization. The council's responsibilities span the entire AI life cycle, from ideation to ongoing evaluation. As an executive, you provide strategic guidance, leaning on your team for detailed execution. The council also plays a crucial role in addressing the needs and concerns of key stakeholders: employees, customers, partners, and the wider community. This approach fosters trust and support among these groups.

With this governance framework in place, it's time to identify the opportunities for AI, which may begin as efficiency measures but evolve into solving previously unresolved problems.

Once leadership is in place, the next challenge is integrating AI into the organization in a way that enhances your employees' productivity rather than disrupting it. In the next chapter, we'll explore the evolving structure of AI-driven organizations, strategies for preparing the workforce, and best practices for embedding AI into business operations.

Takeaways and Tips

1. **AI Leadership requires more than technical expertise.** It's not enough for an AI leader to be a data scientist. They must also be strategic thinkers who understand business priorities, can communicate effectively across teams, and can drive AI adoption in a way that fits with the organization's short- and long-term goals.

2. **Curiosity is essential.** The best AI leaders are lifelong learners who stay ahead of trends, continuously explore AI advancements, and translate them into business opportunities.

3. **Communication and influence are critical.** AI leaders must bridge the gap between technical and non-technical teams. The ability to clearly articulate AI's value, demystify complex concepts, and secure buy-in from stakeholders is as important as technical knowledge.

4. **AI leadership can come from unconventional backgrounds.** Organizations should consider hiring AI leaders with experience in innovation, ethics, behavioral science, or business strategy, as well as those with technical backgrounds. Diverse perspectives can help drive more thoughtful and responsible AI adoption.

5. **Hiring full-time vs. fractional AI Leaders depends on business needs.** Larger enterprises with complex AI initiatives may benefit from a full-time AI leader, while smaller organizations or those in the early stages of AI adoption might find a fractional AI executive more cost-effective and flexible.

Actionable Steps

- **Rethink your job descriptions**. Move beyond traditional AI roles focused solely on technical expertise. Instead, emphasize skills such as strategic thinking, leadership, and ethical AI governance.

- **Look for cross-industry talent**. Consider candidates with experience in business strategy, digital restructuring, or ethics, as well as AI knowledge. Many of the best AI leaders come from nontraditional backgrounds.

- **Determine if you need full-time or fractional leadership**. Evaluate your organization's AI goals, budget, and stage of AI adoption to decide whether a full-time or fractional AI leader is the right fit.

- **Set up an AI governance council**. Assemble a diverse team, including your AI Lead, executive leadership, IT, legal, compliance, and HR or ethics representatives, to strategically oversee AI initiatives. Direct the group to build a guiding document for success.

CHAPTER 8:
Organizational Design
in an AI-Driven World

Embracing AI involves addressing employee concerns, investing in upskilling, and building a workforce capable of collaborating with AI so it boosts, instead of disrupts, employee productivity. Change creates uncertainty, and uncertainty can cause business processes to slow down, leading employees to worry about job security. Left unaddressed, employees may subconsciously sabotage AI pilots to preserve their roles—a scenario your organization must actively avoid. What's more, AI's presence raises important questions about organizational structure: Should AI function as a standalone department, be integrated within existing teams, or adopt a hybrid model that combines centralized strategy with decentralized execution?

In this chapter, you'll learn:

- How AI is changing job roles across industries and what this means for employees.

- Strategies to prepare your workforce for AI-driven change.

- The advantages and trade-offs of different AI organizational structures.

- How to select the right AI tools to enhance productivity and decision-making.

- A step-by-step guide to integrating AI into your company's organizational framework.

AI doesn't replace human ingenuity; it amplifies it. The key to success lies in structuring AI initiatives in a way that enhances efficiency while empowering employees to take on more strategic, creative, and value-driven roles. Whether you're evaluating AI's implications on your workforce or rethinking your company's structure, this chapter will guide you through designing an organization that is future-ready in the age of AI.

The Changing Workforce in the Modern World

AI is fundamentally reshaping work across industries. In data management, for example, AI streamlines data-heavy tasks like fraud detection and loan application reviews, increasing speed and accuracy. Manufacturers use AI-powered robots to handle tasks such as welding and quality control. In customer service departments, chatbots and virtual assistants resolve routine queries, allowing human representatives to focus on complex, personalized interactions.

While some routine jobs may phase out, AI is redefining the work of knowledge professionals by enhancing their capabilities and helping them produce higher-quality work in less time.

A 2024 Boston Consulting Group (BCG) study found that employees with access to AI tools completed 12.2% more tasks, worked 25% faster, and produced output that was 40% higher in quality than those without AI. Notably, AI had the greatest impact on lower-skilled workers, improving their performance by 43%, compared to a 17% improvement among more experienced employees.

This finding suggests that AI can serve as an equalizer, bridging skill gaps and enabling faster upskilling within organizations. Instead of replacing expertise, AI helps professionals across industries make better decisions, analyze complex data, and focus on higher-value work.

Addressing the Disruption of AI on the Workforce

To successfully integrate AI, company leaders need to first clearly address and openly communicate to their staff how AI will affect the workforce.

- **Acknowledge concerns and provide clarity**. Fear is a natural response to change, especially when AI automates tasks employees once performed. Leaders must emphasize that AI is not a replacement for human expertise but a tool designed to enhance productivity and decision-making.

- **Shift the focus to new opportunities**. By handling repetitive, low-value tasks, AI enables employees to focus on strategic, creative, and people-centric work. For example, automating data entry or routine customer service inquiries allows employees to shift toward problem-solving, innovation, and personalized customer engagement.

- **Invest in upskilling**. Organizations must equip employees with relevant skills so that AI can be an enabler rather than a disruptor. Training in areas such as prompt engineering, AI-driven analysis and decision-making, and human-AI collaboration helps employees remain valuable contributors in an evolving workplace.

- **Reinforce the value of human-AI collaboration**. Real-world examples of AI augmenting, rather than replacing, human roles help shift mindsets. Whether AI assists doctors in diagnostics or supports analysts in data-driven decision-making, showcasing these collaborations builds confidence in AI's role as a partner rather than a competitor.

Preparing the Workforce for AI

Companies must take a proactive approach to workforce metamorphosis so employees can adapt and thrive in an AI-powered environment. Your workforce understands your brand deeply and holds invaluable institutional knowledge. Instead of replacing them with AI, transition employees into new roles where their expertise can continue to benefit your organization. This approach leverages existing knowledge while retaining employee loyalty and engagement.

Additional considerations:

- **Redefine roles**. Identify tasks that can be automated and create pathways for employees to shift into more strategic, creative, or customer-focused work.

- **Invest in continuous learning**. Offer upskilling programs in areas like AI-driven decision-making, data analysis, and human–AI collaboration.

- **Integrate AI into hiring and training**. Prioritize candidates who blend technical proficiency with strategic thinking and create AI literacy programs to train existing employees.

- **Build strategic partnerships and collaborations**. For example, companies should explore working with universities to develop AI programs aligned with industry needs and apply for government grants that support AI workforce training.

AI's Impact on Company Structure

Integrating AI into business operations sparks a crucial debate about its placement within the organizational structure. The choice of how to structure your AI initiatives depends on several factors unique to your business:

- **Size and Complexity**: Larger organizations benefit from centralized AI for consistency, while smaller companies may thrive with decentralized or hybrid models that foster agility.

- **Industry Regulations**: Highly regulated sectors, such as healthcare and finance, require centralized or hybrid approaches to maintain strict data control and compliance. In less regulated industries, a more experimental, decentralized approach can be used.

- **Culture**: Companies with a strong culture of experimentation excel with decentralized models that empower individual teams. In contrast, risk-averse organizations might prefer centralized control to achieve stability and uniformity.

- **Resources**: Organizations with ample resources can invest in decentralized or hybrid strategies, enabling independent exploration across units. Those with limited resources may find that a centralized model offers the best return by concentrating talent and budget where it matters most.

With these factors in mind, organizations must decide whether AI should function as a dedicated department or be integrated within existing teams. The following sections explore the pros and cons of both approaches.

AI as a Separate Department

Here are some pros and cons of creating a dedicated AI department.

Pros:

- **Specialization and Focus**: A dedicated AI department provides concentrated expertise and resources. Specialists can focus on developing, deploying, and managing AI solutions without the distractions of other departmental responsibilities.

- **Consistency and Standardization**: Centralized AI departments can establish uniformity in AI tools, methodologies, and practices across the organization, leading to more consistent outcomes and easier maintenance.

- **Strategic Integration**: A central team can develop AI initiatives that meet the company's overall strategy and goals.

Cons:

- **Potential Bottlenecks**: Centralized teams may become overburdened, slowing down the deployment and scaling of AI projects.

- **Lack of Flexibility**: A centralized, approach might not meet the specific needs of different departments or units.

- **Missed Opportunities**: Individuals may wait to be told what to do with AI, which could lead to overlooking creative opportunities for AI integration.

AI Integrated Within Existing Teams

Here are some pros and cons of integrating AI within existing departments.

Pros:

- **Agility and Responsiveness**: Decentralized units can quickly implement AI solutions tailored to their unique challenges.

- **Empowerment and Ownership**: By giving control to individual units, employees are more engaged and invested in AI initiatives.

- **Scalability**: Decentralized approaches can scale more easily as each unit can grow its AI capabilities independently.

Cons:

- **Inconsistency**: Without a central authority, AI implementation can be inconsistent, with disparate systems and potential integration challenges.

- **Duplication of Efforts:** Different units may work on similar projects without coordination, leading to redundant efforts and wasted resources.

- **Lack of Strategic Cohesion**: Without a unified strategy, decentralized AI efforts may not meet the company's overall goals, potentially resulting in fragmented or conflicting initiatives.

Bridging the Gap: The Hybrid Model

A hybrid AI structure balances centralized strategy with decentralized execution. In this model, a small, central AI team sets the overall direction, plus smaller AI groups directly embedded in each department handles the AI execution—the best of both worlds.

Why This Works:

1. **Central AI Team**: This team defines roadmaps, establishes data security, and prioritizes company-wide objectives while departments customize AI solutions for their specific needs.

2. **Embedded AI specialists**: These individuals understand their department's workflows and can implement solutions more effectively than a centralized team alone. Marketing AI experts refine customer targeting, while operations AI teams focus on supply chain efficiency.

3. **Departmental Experts**: Departmental AI teams experiment with ideas, sharing insights with the central team to refine best practices and prevent redundant efforts. Successful pilots can be expanded company-wide with minimal risk.

4. **All: Culture of AI Ownership**: Empowering departments to figure out AI adoption fosters engagement and adaptability, making AI an integrated tool rather than an external initiative.

A well-designed structure, however, is only as effective as the tools that support it. To fully realize AI's potential, organizations must carefully select and integrate AI solutions that suit their workflows, strengthen decision-making, and increase measurable business value.

Where to Begin to Become an AI Integrated Organization

Starting with a top-level view of your organization and ending with a detailed map of tasks ready for AI enhancement can feel overwhelming if you've never done it before. However, breaking down AI adoption step-by-step and utilizing tools like JobsGPT can make the integration process both manageable and insightful.

Created by Paul Roetzer, JobsGPT is a ChatGPT-powered solution that generates detailed task lists for corporate roles and identifies which tasks are well-suited for AI assistance. It helps estimate potential time savings and evaluate the broader impact of AI on job functions, It's a valuable tool not only for streamlining operations but also for reimagining how talent and technology can collaborate. Here's how you can proceed with it:

Step 1: Begin with Your Organizational Chart

The organizational chart illustrates who reports to whom and how responsibilities flow, but it doesn't provide much insight into what people actually do. A job that appears administrative on paper might involve significant decision-making or client interaction in practice, while a senior strategic role might include a surprising number of routine, repeatable tasks. Without digging into these details, it's difficult to identify which responsibilities could be augmented, streamlined, or even reimagined with AI tools.

When evaluating where AI could make a meaningful impact, it's important to look beyond titles and reporting lines and focus instead on the actual tasks being performed.

Action: Begin by identifying one or two departments or functions that are strong candidates for AI integration—perhaps those that are data-rich, process-heavy, or customer-facing. These could include your marketing, operations, or customer service teams.

The following example focuses on a supply chain management role. I chose this role because it encompasses a broad range of responsibilities—some that require uniquely human skills, such as relationship-building and negotiation, and others that could be significantly streamlined with AI tools, such as demand forecasting or logistics coordination.

On the next page is a table of the tasks, exposure level (explained in Step 3), estimated time saved by utilizing AI in the supply chain role, and the rationale behind the estimates.

Supply Chain Management Role Analysis

Task	Exposure Level	Estimated Time Saved (%)	Rationale
Demand forecasting	E1/E6	30-50%	AI can analyze historical data, market trends, and external factors to predict demand. Advanced reasoning helps optimize accuracy.
Supplier relationship management	E0	0%	Managing relationships and negotiating with suppliers typically requires human interaction and emotional intelligence.
Inventory management	E1/E2	20-40%	AI can optimize inventory levels by analyzing sales patterns and suggesting reorder points. AI-powered tools can automate reordering processes.
Logistics coordination	E2/E8	30-50%	AI integrated with logistics software can plan and route shipments efficiently, track deliveries, and ensure optimal cost management.
Procurement strategy development	E1/E6	30-50%	AI can analyze market conditions, supplier performance, and internal needs to support the creation of procurement strategies, improving decision-making quality.
Cost analysis and reduction	E1/E6	20-40%	AI can analyze procurement costs, identify inefficiencies, and suggest cost-saving opportunities with advanced data analysis capabilities.

Step 2: Gather Role Descriptions

Once you've chosen a department, collect the job descriptions for each relevant role. If the descriptions are outdated, consider hosting short interviews with team leads and employees to confirm their daily tasks. The goal is to fill in the details the organizational chart can't provide, such as actual responsibilities, routine tasks, and critical workflows.

Action: Assemble current job descriptions and refine them with input from those who do the work. Aim to break down each role into a list of specific tasks rather than broad responsibilities. For example, "Developing marketing strategies for oncology services" might involve brainstorming, data research, competitive analysis, and proposal drafting.

Step 3: Identify Tasks and Their "Exposure" to AI

With a granular task list, you can now consider how AI might help. This is where the "Exposure Key" comes into play. The Exposure Key defines what level of AI capabilities each task could manage—from basic language support (E1) to more advanced reasoning (E6) or even digital action (E8). As you evaluate each task, ask: "If I applied AI, what type would I need?" For instance, would a large language model (LLM) suffice, or would advanced reasoning or digital-world action capabilities be beneficial?

Action: For each task, assign an "Exposure Level" based on the Exposure Key. Something like "drafting marketing strategies" might be E1 (direct exposure to LLM capabilities), while "analyzing complex market trends and drawing strategic conclusions" might be E1/E6 (since it benefits from both basic LLM input and advanced reasoning).

Exposure Key:

E0	No exposure
E1	Direct exposure
E2	Exposure by LLM-powered applications
E3	Exposure given image capabilities
E4	Exposure given video capabilities
E5	Exposure given audio capabilities
E6	Exposure given advanced reasoning capabilities
E7	Exposure given persuasion capabilities
E8	Exposure given digital world action capabilities (AI Agents)
E9	Exposure given physical world vision capabilities (AI Vision Devices)
E10	Exposure given physical world action capabilities (Humanoid Robots)

Source: Smarterx.ai

Step 4: Use Tools like JobsGPT by SmarterX

To streamline this process, you can turn to specialized AI analysis tools. JobsGPT from SmarterX is free to use and can take your refined job descriptions and tasks and then provide insights on which activities are ripe for AI integration. It can estimate time saved and suggest which levels of exposure are most relevant.

Action: Input your task breakdown and responsibilities into JobsGPT. Review the output, which will detail the time-saving potential and the nature of the AI support (e.g., automated content drafting, event coordination, data analysis).

Step 5: Assign AI "Spots" in the Org Chart

As you evaluate which tasks can be supported or automated by AI, given its high capability and autonomy, consider how AI "agents"—intelligent tools that can act independently to execute complex tasks and workflows— might fit into your organizational structure. In the future, you could have a "Digital Marketing Agent" that assists multiple people by drafting social content or analyzing campaign performance. This "agent" essentially occupies a position on your org chart, coordinated and supervised by a person, until it is advanced enough to operate more independently.

Action: Identify where AI agents could sit alongside team members. For instance, in a marketing department, a content generation agent could report to the content manager, who provides oversight, quality checks, and strategic direction.

Step 6: Prioritize

Not all tasks need immediate AI intervention. Start with those that promise the greatest performance gains or strategic insights. Run pilot programs, train your team to use these AI tools, and evaluate the outcomes. As confidence and competence develop, you can address more complex tasks or introduce additional AI agents with greater exposure levels.

Action: Implement a pilot for a select group of tasks. For example, begin with automating certain reporting functions (E1/E2 tasks) before moving on to more complex analytical tasks (E6).

Step 7: Scale and Continuously Refine

As you experience success, gradually expand into other roles and departments. Revise your job descriptions and organizational structure to reflect the new capabilities. Over time, your organizational chart may evolve to showcase a blend of people-led roles and AI agents working in harmony. Continuously reassess and remap tasks as technology advances and your workforce becomes more AI-savvy.

Action: Periodically revisit your task analysis and exposure levels. AI tools that were once experimental may now be standard. Adjust your organizational design to integrate these AI roles more seamlessly.

By following these steps and leveraging a tool like JobsGPT, you're not just layering AI on top of your existing structure; you're integrating it into the fabric of how work gets done. Over time, this careful, methodical approach will help you build an organization that thrives in an AI-driven world.

The next step in AI adoption is to move from strategy to execution. How can organizations test AI's potential in real-world scenarios? The next chapter will explore AI use cases and pilot programs, providing a roadmap for effectively launching and scaling AI initiatives.

Takeaways and Tips

1. **AI integration requires a strategic approach**. AI fundamentally reshapes how businesses function, and organizations must align AI initiatives with long-term business objectives to gain a competitive advantage.

2. **Workforce adaptation is critical for AI success**. AI's value is maximized when employees are equipped to collaborate with it. Companies must invest in upskilling and reskilling programs to help workers transition into AI-enhanced roles and develop essential AI literacy.

3. **Choosing the right AI organizational model matters**. Your company's size, industry, resources, and agility needs will determine whether you choose a centralized, decentralized, or hybrid AI model. The hybrid approach often effectively balances control and responsiveness.

4. **AI adoption is an evolving process, not a one-time initiative**. AI capabilities, regulations, and best practices continuously evolve. To stay competitive, organizations must embrace adaptability, foster a culture of continuous learning, and refine AI strategies over time.

5. **Start small, scale strategically, and refine**. Begin AI integration by targeting specific departments and tasks, refine your approach through pilot programs, and then scale successes across the organization, continuously updating your approach as AI matures.

Actionable Steps for AI-Driven Organizational Design

- **Assess employee tasks and AI opportunities.** Identify departments or roles most suitable for AI integration. Collect current job descriptions and engage employees to identify repetitive, data-driven, or process-heavy tasks suitable for AI augmentation.

- **Communicate transparently and proactively.** Host open forums or Q&A sessions to discuss AI's role, address fears, and clarify misconceptions. Emphasize AI as a tool that augments rather than replaces people.

- **Launch targeted upskilling programs.** Create training programs focused on AI literacy, prompt engineering, AI-assisted analytics, and human-AI collaboration. Partner with universities or specialized training providers to bring in learning initiatives for practical business needs.

- **Determine your optimal AI operating model.** Based on your company's size, complexity, regulatory environment, and resources, evaluate whether your AI structure should be centralized, decentralized, or hybrid. Clearly outline how AI-related roles fit into your existing organizational structure.

CHAPTER 9:
Identifying AI Test Cases and Tools

The most successful AI initiatives begin with a strategic approach to integrating AI into existing business models. However, many organizations face a fundamental question: Where should we start? Without a clear framework for identifying AI use cases, companies risk investing in pilots that fail to deliver.

In this chapter, you'll learn:

- How to shift the focus from technology-first to business-first AI adoption.

- The three strategic areas where AI delivers the most value: efficiency, quality, and innovation.

- How to identify AI-ready processes within your organization.

- A structured framework for evaluating AI pilot projects.

- The role of cross-functional collaboration in successful AI implementation.

- How to avoid common pitfalls when launching AI pilots.

- How to build your AI Toolbox.

While the allure of flashy AI projects is strong, successful AI implementation is rooted in careful consideration and targeted use cases. In this chapter, we'll guide you through the process of pinpointing AI's highest-value opportunities and structuring pilot programs for success.

Start with Objectives, Not Technology

The cornerstone of successful AI adoption is clear harmony with business objectives. AI is not a magic wand; it is a tool designed to enhance and accelerate existing processes. Instead of asking, *"Where can we use AI?"* begin with, *"What are our critical business goals?"*

For instance, if your organization prioritizes improving customer satisfaction, consider chatbots to enhance response times or personalized recommendation systems to deepen engagement. If operational excellence is a key goal, AI can predict maintenance needs or optimize supply chain logistics. By letting strategic objectives guide AI adoption, technology becomes a means to an end rather than an end in itself.

Identifying High-Impact AI Use Cases: Three Strategic Focus Areas

In the following sections, we'll examine three core strategic focuses where AI can make a transformative difference:

1. Increasing Efficiency

2. Improving Quality

3. Innovating Products and Services

Each of these areas demonstrates how AI solves problems and creates new opportunities for growth and value.

1. Increasing Efficiency

Many businesses begin their AI journey by focusing on efficiency. They use automation to eliminate mundane tasks, enabling teams to concentrate on higher-value work. This involves distinguishing between "thinking" tasks, which require critical judgment, and what I call "thunking" tasks, which are repetitive or routine. AI excels at automating these "thunking" activities, enhancing workforce productivity.

Identify repetitive, time-consuming tasks that don't require deep human insight; these are ideal candidates for AI automation, which can free up resources for strategic initiatives. AI also provides scalable solutions for business processes that struggle to keep pace with rising demand, ensuring smoother operations and improved customer experience satisfaction.

Case Study: Toyota's Just-in-Time (JIT) inventory management system highlights remarkable efficiency by reducing inventory levels by up to 60% and cutting carrying costs by 20% to 40%, while maintaining only four hours' worth of inventory at its Georgetown, Kentucky, plant. These tangible results demonstrate how strategic, data-driven approaches optimize supply chains and boost operational performance.

2. Improving Quality

AI's ability to enhance quality extends across multiple dimensions, including accuracy, consistency, and decision-making. By performing repetitive tasks with high precision, AI significantly reduces human error, making it invaluable in fields that require meticulous data analysis or complex calculations.

Beyond accuracy, AI-driven tools analyze vast datasets to uncover valuable insights that might otherwise go unnoticed, speeding up decision-making and ensuring strategies are grounded in comprehensive, real-time information. AI also boosts predictive capabilities by examining historical data and identifying patterns, allowing businesses to anticipate trends, forecast demand, and proactively tackle potential challenges before they escalate. In customer-facing sectors such as marketing and service, AI further enhances quality by personalizing interactions based on individual behaviors and preferences, promoting stronger engagement and improving overall client satisfaction.

Case Study: The Swedish-based fintech company Klarna implemented an AI assistant to handle customer service chats, successfully managing two-thirds of interactions within the first month. This not only streamlined customer support and reduced response times but also enhanced issue resolution accuracy, resulting in a 25% decrease in repeat inquiries and reducing average resolution time from 11 minutes to under 2 minutes.

3. Innovating Products and Services

AI is helping businesses develop and refine products, especially in niche markets abundant with customer data. By analyzing market trends, customer preferences, and competitors, AI assists executives in identifying unmet needs and emerging opportunities. The result? New products remain relevant and strategically positioned. AI's predictive capabilities also enable businesses to anticipate consumer behavior and industry shifts, facilitating proactive decision-making in product design.

Integrating AI into existing products improves functionality and user experience, making offerings more intuitive and responsive. Features like personalized recommendations or predictive maintenance can set products apart in competitive markets. Plus, AI can automate complex processes and help create solutions previously unattainable, expanding possibilities for groundbreaking products that meet specialized market demands.

Case Study: Blue River Technology is changing the farming game with its smart See & Spray system. Using artificial intelligence, it can quickly tell the difference between crops and weeds, allowing farmers to spray only where needed. This cuts herbicide use by up to 90%, saving money and protecting the environment at the same time. Now part of John Deere, Blue River continues to create sustainable solutions that make farming easier and more efficient.

Identifying Opportunities

Choosing the right AI pilot programs begins with thoroughly understanding how work is conducted throughout the organization. While leaders may establish strategic priorities, the most valuable AI opportunities often arise from the workflows and challenges employees face daily.

To uncover these insights, consider a multi-week listening and observation tour where your AI leader is immersed in various departments, observing processes and tracking projects from beginning to end. This hands-on approach confirms that AI pilots are based on actual operational needs rather than assumptions. Here are some important points to keep in mind for this exercise:

1. **Be comprehensive**. Don't hold back. Remove all judgment and jot down every idea that comes to mind. Think about the challenges or opportunities that your company has faced.

2. **Break it down**. It is often easier to think about AI integration in your processes or roles, not as a single end-to-end solution, but as individual components that fit together like pieces of a puzzle. For example, in employee onboarding, AI might help generate a two-week onboarding plan or answer simple questions for a new employee, but the human element remains essential for fostering connection and ensuring a smooth transition.

3. **Look ahead**. Document the opportunities even if current AI capabilities fall short of achieving them today. For instance, you may have wanted to use AI to identify speech recognition patterns in pitch and intonation, in addition to simple sentiment analysis for customer support calls. At one point, the technology wasn't available, but nine months later, Hume.AI released an interface that did just that, focusing on emotional intelligence analysis. This example highlights the importance of documenting ideas, as AI capabilities are rapidly evolving, and a solution that isn't feasible today may soon become achievable.

4. **Collaborate**. Be ready for discussions with colleagues to fully complete the framework. You might not have all the necessary details to provide a value creation statement or to estimate the hours needed for resolution. By engaging with others, you'll help bridge these gaps and complete a more thorough assessment. Start with a governance council to lay the foundation for your AI goals (refer to Chapter 7), but it is essential to involve a broader group of employees who interact with both internal and external stakeholders. This wider engagement will offer diverse perspectives and insights that are critical for success. We will delve into this more when we talk about getting people on board.

Evaluating AI Pilot Programs

Once potential AI opportunities are identified, you should evaluate pilot programs to assess whether they deliver practical value and meet strategic goals. A structured assessment framework helps balance innovation with feasibility, preventing organizations from being sidetracked by unproven technologies.

This framework offers a comprehensive method for assessing and prioritizing AI initiatives, making it especially valuable for reporting to boards or executives. By systematically analyzing each potential AI project across various dimensions—such as estimated value, implementation time, likelihood of success, and impact on change management—decision-makers can make informed choices about resource allocation. The framework promotes a portfolio approach, balancing quick wins with more ambitious "moonshots" that hold the potential to transform business.

The AI Framework

When I first started brainstorming AI business application ideas, I kept a notebook filled with concepts. However, I often found myself overwhelmed, unable to prioritize, and struggling to get executive buy-in. Then, at the Marketing AI Conference, I discovered a framework that provided a structured way to prioritize pilot programs and effectively communicate them to executives or boards. Over the past two years, I've refined this framework further through real-world use.

Below, I outline the framework's components, providing a step-by-step guide and an example to show how each part is implemented. You can download a customizable version of the framework at https://www.demystifai.com/aimadesimple/execlibrary. Feel free to adapt it to suit your organization's needs.

1. AI Solution Category

Define the pilot's objective by determining its focus. This might be to improve output, increase revenue, reduce churn, enhance decision-making, or address another specific challenge. While this might seem like a general business strategy, that's precisely the point. AI should always serve strategic business goals and not exist as an isolated initiative.

Example: *Revenue Growth*

2. Challenge or Opportunity Statement

Clearly articulate the business challenge, inefficiency, or market gap the pilot addresses. Quantify the value of the challenge and identify key stakeholders. Explain how the pilot fits with organizational goals and its potential benefits.

Example: *Our business development team of three only has the bandwidth to complete four RFPs annually despite each being a high-value contract exceeding $1 million.*

3. Value Impact Statement

Highlight the tangible benefits AI will deliver, linking them to the organization's strategic goals. This is critical for gaining executive or board approval, as it illustrates the initiative's rationale and projected value.

Example: *Currently, we win 50% of RFPs. By leveraging AI, we could complete six RFPs annually, potentially securing three additional clients per year.*

4. Estimated Year 1 Value

Calculate the anticipated revenue or savings for the first year of implementation. Estimate this value using historical data or average costs and time spent on relevant tasks.

Example: *If three new clients are secured with average contract values of $1 million and contracts begin mid-year, the value created in Year 1 would total approximately $1.5 million.*

5. Total 3-Year Estimated Value

Forecast the cumulative value over three years, accounting for scaling and growth opportunities. Identify whether the pilot offers exponential value as it matures.

Example:

Year 1: $1.5 million (contracts begin mid-year)

Year 2: $3 million

Year 3: $6 million

Total: $10.5 million

6. Solution Type

Classify the AI pilot as a quick win, sweet spot, or moonshot:

- **Quick Wins**: Low complexity, moderate impact, immediate benefits.

- **Sweet Spots**: Medium complexity, high impact, aligned with key pain points and goals.

- **Moonshots**: High complexity, groundbreaking, significant investment, potential to redefine business models.

Example: *Sweet Spot*

7. Hours to Solve

Estimate the time required for implementation, from data preparation to deployment. Use a T-shirt sizing approach for flexibility:

- XS (10–50 hours): Minor data analysis or simple tasks

- S (50–150 hours): Small-scale projects like a chatbot

- M (150–400 hours): Moderately complex tasks

- L (400–800 hours): Multi-department initiatives

- XL (800+ hours): Enterprise-level systems

Example: *Medium (~275 hours)*

8. Probability of Desired Outcomes

Assess the likelihood of success by evaluating the following:

- **Data Quality**: Ensure data is accurate, complete, and representative.

- **Technical Feasibility**: Confirm the tools, infrastructure, and expertise are available.

- **Organizational Readiness**: Gauge stakeholder buy-in, team skills, and capacity for change management.

Example: *75% likelihood of success, as increasing RFP output may not maintain the same win rate.*

9. Change Management

Estimate the scope of change required, categorized as low, medium, or high impact. Consider the number of departments or processes affected and balance this with overall organizational readiness to avoid change fatigue.

Example: *Low impact, limited to the Business Development team.*

10. Potential Roadblocks

Anticipate and address challenges such as legal compliance, data quality, security risks, or ethical concerns.

Example: *Limited access to relevant data due to the absence of a centralized RFP repository.*

11. Cost

Detail the costs of the pilot, including technology investments, personnel, and external consulting fees. Conduct ROI and cost-benefit analyses to justify the expenditure.

Example: *$15,000*

Pilot Projects: Start Small, Scale Fast

1. **Start with small, impactful pilots.** Begin by testing AI on a manageable scale. Choose projects that are both feasible and significant, ones that quickly demonstrate value but don't require overhauling your entire infrastructure. This approach lets you control risks, gather data, and learn valuable lessons before a full-scale rollout.

2. **Define success with clear metrics.** Set measurable goals that tie directly to business objectives. For example, a pilot might aim to reduce processing time by 30% or boost customer satisfaction by 15%. In today's digital age, your organization likely already collects data that can be leveraged to measure the impact of AI pilots. Short timelines (a few months) help you get rapid feedback and maintain momentum. Involving key stakeholders across departments also builds ownership and uncovers potential hurdles early.

3. **Embrace learning from failure.** Not every pilot will hit the mark, and that's okay. Use any setbacks as learning opportunities. Document both successes and challenges to refine your approach, ensuring that each pilot, successful or not, provides insights for future projects.

4. **Scale what works.** Once a pilot proves successful, plan to expand the solution across the organization. Key steps include:

    *   **Change Management**: Engage internal champions who can advocate for the benefits of AI and help ease transitions.

- **Infrastructure Readiness**: Upgrade systems as needed to support broader AI deployment.

- **Ongoing Monitoring**: Set up feedback loops to continually assess performance and adapt to evolving business conditions.

5. **Keep your AI initiatives organized**. Maintain a master document of all AI pilot projects. Regular reviews help reprioritize initiatives based on new technological advancements and changing business needs, keeping your strategy agile and forward-thinking.

6. **Leverage external partnerships**. Partnering with experienced AI vendors or consultants can accelerate pilot deployment and scaling. These partners can fill gaps in internal expertise, provide pre-built models and proven workflows, monitor compliance with regulations and manage risks, and offer access to the latest technology. They can also help match AI solutions to your strategic goals.

The following framework organizes your AI initiatives and helps to clearly communicate their value to stakeholders. By grounding your AI projects in measurable business outcomes, you can build trust, secure buy-in, and produce meaningful results for your organization.

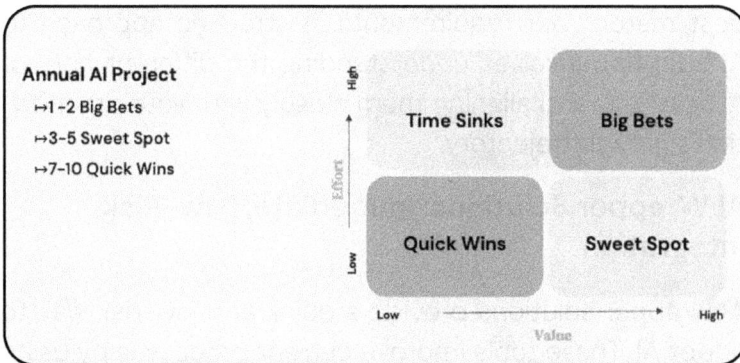

As you can see in the chart, I have included my recommendation on what most companies should strive to take on in a year:

- 1-2 Big Bets

- 3-5 Sweet Spot

- 7-10 Quick Wins

I would halve this for the first year, as you will be developing policies and setting up infrastructure for your governance and cross-departmental councils.

Building Your AI Toolbox: Identifying the Right Tools

Your AI pilots have provided critical insights into where AI delivers the most value in your organization. The next step is selecting the right AI tools to scale these successes. It's not about simply choosing just one AI tool from a list while ignoring others; rather, as you progress through your AI pilots, you'll begin to clearly identify the features and capabilities your organization truly needs. This process naturally guides you toward selecting the AI solutions that best match your requirements. A strategic approach to AI adoption involves understanding the different tiers of AI solutions and aligning them closely with your organization's growth trajectory.

AI Wrapper Solutions: Immediate, Low-Risk Integration

AI wrapper solutions provide a quick and low-risk way to adopt AI. These tools improve current processes by using foundational AI models without needing major system overhauls. They vary from straightforward AI-powered automation upgrades to generative AI integrations added to existing workflows.

- Quick to implement and cost-effective.

- Ideal for organizations looking to experiment with AI while minimizing risk.

- Common use cases include AI-powered customer support, automated content generation, and workflow optimization.

For instance, Julius.ai acts as a wrapper around large language models like ChatGPT or Claude, making them easier for businesses to use effectively. Julius.ai adds extra functionality that the core AI model itself doesn't offer directly, such as document management, team collaboration, integration with company-specific data, or streamlined workflows. By doing this, Julius.ai converts a general-purpose AI into a specialized tool customized for real-world business applications.

Gen-AI Native and Narrow Solutions: Targeted, High-Impact Applications

These purpose-built AI tools are designed to solve specific business challenges. They offer high accuracy and specialized functionality. While they require more technical integration, their precision and effectiveness make them invaluable.

- Best suited for solving critical business pain points.

- Often require deeper integration and AI expertise.

- Common use cases include fraud detection, predictive maintenance, and personalized marketing AI.

A good illustration is Writer.ai, a focused AI platform dedicated to improving how businesses create, edit, and manage written content efficiently. It uses AI to help businesses generate high-quality, on-brand written content, enforce consistent brand voice, optimize copy for SEO, and improve overall communication clarity. Unlike general-purpose AI platforms, Writer.ai is finely tuned to handle the singular task of writing optimized content exceptionally well, ensuring targeted effectiveness.

Core Platforms with New AI Functionality: Enhancing What Already Works

Many enterprise software providers are integrating AI features into their existing platforms, making it easier for businesses to adopt AI without disrupting familiar workflows. This approach enables companies to gain advanced AI capabilities while maintaining user-friendly interfaces.

- Provides a balance between innovation and ease of adoption.

- Reduces disruption by keeping AI within existing systems.

- Common use cases include AI-driven analytics, process automation, and personalized customer interactions.

Salesforce is a great example of a core platform that's enhanced its value by integrating new AI features into existing workflows. Rather than introducing entirely new tools, Salesforce added Einstein AI directly into its CRM, helping sales teams better identify leads, predict customer behavior, and automate routine tasks. It also gives customer service teams personalized insights and sentiment analysis, allowing them to proactively resolve issues and improve customer experiences without leaving the familiar Salesforce environment.

Core Operational Systems: Transformative AI at the Core

At the highest level of AI adoption, organizations fully integrate AI into their core operational infrastructure. This approach enables informed decision-making, automation at scale, and continuous optimization.

- Requires significant investment and organizational change.

- Delivers the highest shift in efficiency, agility, and innovation.

- Common use cases include supply chain management, real-time dynamic pricing, and autonomous
decision-making.

Consider Google, which offers not only Cloud services
but also integrates its robust, scalable infrastructure with
powerful AI models like Gemini. This enables businesses
to seamlessly incorporate cutting-edge AI into their core
operations. Such a tight integration empowers companies
to optimize performance, enhance decision-making, and
innovate rapidly, all within a unified and reliable platform.

Executives should combine their AI strategy with current
capabilities and future ambitions. For quick wins, start
with AI wrapper solutions, then gradually integrate more
specialized AI tools. When ready for broader AI integration, leveraging AI-enhanced platforms can facilitate the
transition. Core operational AI will ultimately redefine the
organization's competitive edge.

Picking Your Core AI Foundational Tool

Your core AI platform serves as the backbone of your
AI strategy. This is the tool you commit to for the long
term, ensuring consistency and deep integration across
business functions.

- Stick with one platform. Just as organizations
standardize key software, it's essential to commit
to a single AI platform for company-wide adoption.

- Recommended platforms include OpenAI and
Google Gemini, which offer extensive capabilities
such as advanced data analysis and generative AI.

- Although Microsoft Copilot integrates seamlessly

into the Microsoft ecosystem, its overall performance hasn't matched up with other AI solutions. While many organizations use it because it's convenient and cost-effective within Microsoft 365, feedback has been mixed—most praise comes from its Azure and GitHub integrations. In contrast, features like the AI writing assistant in Outlook often fall short, sometimes discouraging hesitant employees from fully embracing AI. Ultimately, successful AI adoption hinges on choosing the right tools that build confidence and deliver consistent value.

By selecting a robust AI platform and investing in an annual contract, organizations gain stability and develop deep familiarity with the tool, resulting in smoother implementation across teams.

Adding Flexibility at the Frontline

While your core AI tool provides a foundation, additional AI tools enable agile adaptation and experimentation. These ancillary tools add targeted functionality without a long-term commitment. Their flexibility allows one to adapt to these changes without disrupting the core AI strategy.

In the next chapter, we'll examine how AI is transforming business models and redefining competitive advantage for a future that is unfolding right now.

Takeaways and Tips

1. **AI adoption should align with business objectives.** Organizations should focus on AI use cases that directly contribute to strategic goals such as operational improvements, quality control, or product development. Starting with a clear business need ensures AI investments deliver measurable value.

2. **AI pilot programs should be structured for impact and scalability.** Successful AI adoption begins with small, well-defined pilot projects that can demonstrate quick wins and build organizational confidence. AI initiatives remain results-driven when pilots with clear success metrics are prioritized.

3. **Cross-functional collaboration is essential for AI implementation.** AI affects multiple business functions, making it critical to involve teams from IT, operations, finance, and customer experience. With a collaborative approach, AI solutions can be integrated more seamlessly across the organization.

4. **Selecting the right AI tools depends on business maturity and readiness.** Organizations should match AI solutions to their level of AI maturity. Start with AI wrappers for quick wins, move to specialized tools for targeted challenges, and eventually integrate AI into core operational systems.

5. **AI success depends on continuous evaluation and iteration.** AI requires ongoing assessment, adaptation, and refinement. Regular reviews of AI performance, evolving capabilities, and emerging use cases will keep AI initiatives compatible with business needs.

Actionable Steps for AI-Driven Organizational Design

- **Listen and observe**. Conduct department-level tours to observe operations and identify AI opportunities directly from employee insights.

- **Establish a structured evaluation of AI pilots**. Implement a structured framework for prioritizing AI pilot projects based on potential business value, implementation feasibility, and organizational readiness. Find a helpful template at **https://www. demystifai.com/aimadesimple/execlibrary**.

- **Build your AI toolbox**. Evaluate the key AI platform tools using an established rubric, such as the one at **https://www.demystifai. com/aimadesimple/execlibrary.** Choose a core, stable AI platform supported by flexible ancillary AI tools on a month-to-month basis.

- **Define clear metrics**. Establish clear, measurable goals for each AI pilot to track success (e.g., improved response times, reduced costs).

- **Conduct quarterly AI strategy reviews**. Then adjust strategy based on technological advances and evolving business needs.

- **Develop external AI partnerships**. Collaborate with external AI experts through partnerships to speed up implementing and scaling your pilot, particularly when internal resources or capabilities are limited.

SECTION THREE:

The Strategic Horizon: From AI Adoption to Industry Transformation

CHAPTER 10:
The Next Evolution in Successful Business Models

Companies that view AI as merely a tool for small improvements may gain short-term benefits, but those that integrate AI into their core strategy will lead the next era of business. Genuine transformation occurs when companies go beyond incremental advancements and use AI as a catalyst for new business models, enhanced customer experiences, and industry disruption.

This chapter examines how companies can move from viewing AI as an enhancement tool to embracing it as a central way to propel breakthroughs and create long-term value. We'll cover the risks of stagnation, the distinctions between AI-adapted and AI-native businesses, and the emerging business models shaping the AI-first landscape economy.

In this chapter, you'll learn:

- Why efficiency-focused AI strategies lead to diminishing returns.

- How leading companies use AI to create new value propositions and streamline existing ones.

- The shift from fixed pricing models to AI-powered outcome-based pricing.

- How AI is moving businesses from one-time sales to continuous value delivery.

- The emergence of smart ecosystems that are replacing traditional industry boundaries.

- The difference between AI-adapted businesses and AI-native businesses, and why it matters.

Companies that fail to recognize AI as a platform for building entirely new business models will find themselves competing in industries that no longer operate under the same rules.

The Challenge of Stagnation

Concentrating solely on efficiency-focused AI initiatives comes with risks. Organizations often experience a plateau where additional improvements yield diminishing returns. Meanwhile, competitors adopting similar AI optimizations rapidly close the gap, neutralizing any initial edge.

Three key risks emerge:

- **Commoditization**: As AI tools become more standardized and accessible, operational excellence shifts from being a competitive differentiator to a baseline expectation.

- **Replicability**: AI-driven process improvements can be easily copied, eroding any temporary edge.

- **Missed Strategic Opportunities**: A narrow focus on internal optimizations can blind organizations to market shifts, emerging customer needs, and industry disruptions.

Identifying these risks is essential for executives seeking stakeholder support. Stagnation signals the need to shift from small improvements to a bold AI-driven vision. Consider two CEOs who begin with similar AI investments but take different approaches.

CEO A celebrates a 15% increase in manufacturing performance. Costs are down, production is up, and the team considers this a significant victory. However, CEO A concentrates solely on enhancing current operations without seeking new opportunities. A year later, competitors have matched these improvements, eliminating the advantage. Once regarded as an innovator, CEO A's company now finds it difficult to stand out in a competitive market.

CEO B, on the other hand, also achieves a 15% efficiency boost but views this as a foundation for real change. Asking bold questions, CEO B considers:

- How can this capability enable entirely new customer solutions?

- Could the company transition from selling products to offering performance guarantees?

- Can AI insights support predictive maintenance or partnerships that redefine the industry?

Using performance gains as a springboard for innovation, CEO B reimagines the company's value proposition, explores new revenue streams, and secures a leadership position in the market. A year later, CEO B's company thrives while competitors struggle to copy its success.

This story illustrates that efficiency alone isn't the goal. AI's real power is in reshaping industries at their core, far beyond process optimization.

Beyond Efficiency: What an AI-First Business Model Looks Like

An AI-first business model necessitates a fundamental mindset shift, moving beyond conventional revenue models, fixed pricing structures, and linear operations. Companies that embrace this transformation are reshaping their industries.

Key Shifts in Business Model Strategy

Three key transitions define how AI-first businesses diverge from traditional models:

1. **From Fixed Pricing to Outcome-Based Pricing:** Traditional pricing models depend on fixed costs, whether through one-time purchases, subscription fees, or tiered service plans. These models assume that all customers will derive the same level of value from a product or service, regardless of individual circumstances. AI enables a shift toward outcome-based pricing, where customers pay for measurable results rather than simply for access to a product or service.

   In an AI-driven outcome-based model, businesses align their revenue with customer success. Instead of charging a flat annual fee for cybersecurity software, an AI-first security provider might price its services based on the number of cyberattacks successfully prevented. Similarly, instead of selling an AI-powered logistics system at a fixed cost, a supply chain company might charge based on the percentage of inefficiencies eliminated.

Outcome-based pricing transcends how businesses charge for their services to how they define their value. AI allows for measuring performance with precision, enabling companies to guarantee specific outcomes and construct their pricing models accordingly.

2. **From One-Time Sales to Continuous Services:** The traditional product-based business model relies on one-time transactions: Customers buy a product, and the relationship largely ends at the point of sale. Even service-based businesses often operate under a transactional mindset, delivering a service once and then moving on to the next client. AI challenges this model by enabling continuous value delivery, where companies remain engaged with customers over time, improving their experience and outcomes through on-going insights powered by AI.

   AI is driving a shift from one-time sales to ongoing service models. Manufacturing companies now offer guaranteed performance instead of selling equipment, using AI to predict and prevent failures. In healthcare, providers are moving from episodic visits to contin-uous AI-powered monitoring, tracking real-time data for personalized care.

   By embedding AI into the core of their offerings, companies can create long-term relationships with customers rather than relying on one-time transac-tions. Instead of simply selling a tool or service, they provide an evolving solution that adapts and improves over time.

3. **From Independent Operations to AI-Powered Ecosystems:** Traditional business models assume that companies operate in isolation, competing within rigid industry boundaries. AI-first businesses, however, recognize that data, intelligence, and automation can be shared across networks, creating interconnected ecosystems that are more valuable than any single company's offering.

An AI-powered ecosystem brings together partners, customers, suppliers, and even competitors who all contribute to and benefit from shared insights. Companies that successfully build these ecosystems can scale exponentially by leveraging collective intelligence rather than relying solely on their own data and capabilities.

This shift is already reimagining industries. Tesla, for example, is creating an autonomous driving ecosystem where owners can rent their self-driving cars to a shared network. More than just an online retailer, Amazon uses AI-powered recommendation algorithms to fuel an entire third-party marketplace. In financial services, AI-powered platforms are replacing traditional investment firms by aggregating real-time market data and providing continuously optimized investment strategies.

AI-powered ecosystems create competitive advantages that are difficult to replicate. While competitors may be able to adopt similar AI capabilities, they cannot easily replicate the network effects of a well-established AI ecosystem. These ecosystems become self-reinforcing, growing stronger as more participants contribute data and interactions, further improving AI's capabilities.

In the AI-first economy, the companies that thrive won't just be those using AI to perform traditional tasks more efficiently. The real winners will be those who leverage AI to fundamentally rethink what's possible. AI offers the flexibility to continuously innovate your products and services and the agility to scale quickly as your customer base expands.

AI as a Market Maker: When AI Creates Entirely New Industries

AI makes it possible to develop products and services that could not have existed before. From synthetic media and autonomous decision-making systems to AI-driven marketplaces, new industries are emerging at the intersection of data, intelligence, and automation. Companies that identify these opportunities early will not only gain a competitive advantage but will define the next decade of market dominance.

The following examples highlight how AI is acting as a market maker, generating new industries that did not exist before its capabilities became viable.

AI-Generated Content Markets

The rise of AI-generated media is creating a new economy for digital content. AI-powered tools can now generate images, music, voiceovers, podcasts, social media posts, videos, and written content with near-human quality, eliminating many traditional production constraints. Businesses and individuals no longer need to rely on human designers, writers, or voice actors to produce content at scale.

This shift has created new marketplaces where AI-generated media is bought, sold, and customized. AI-powered design platforms enable users to create logos, marketing materials, and even entire websites without human intervention. Text-to-image AI models allow brands to generate custom product photography without a photo shoot. AI-generated voice synthesis is disrupting traditional voiceover work, allowing companies to produce branded audio without hiring a professional voice actor.

The implications of AI-generated content extend to gaming and entertainment, where AI is being used to create procedural environments, synthetic actors, and adaptive storylines. In education, AI-generated training materials can be tailored in real time to meet the needs of individual learners. These applications form the foundation of entirely new business sectors centered around AI-generated creativity.

AI-Driven Legal Research and Automated Case Law Analysis

The legal industry has traditionally operated on a billable-hour model, where firms charge clients for the time spent researching case law, drafting documents, and preparing arguments. AI is dismantling this model by introducing instant, automated legal research and document analysis.

AI-powered legal platforms can scan vast databases of case law, statutes, and legal precedents in seconds, providing attorneys with highly relevant insights without the need for manual research. Some platforms even generate legal arguments and draft contracts, reducing the need for human intervention in routine legal work. This has led to the emergence of AI-powered legal research services that charge based on access to automated insights rather than billable hours.

As AI continues to advance, new categories of legal services are likely to emerge. Companies that provide AI-powered legal compliance monitoring, real-time contract analysis, and automated dispute resolution could reshape how businesses and individuals interact with the legal system. In addition to optimizing traditional law firms, these advancements will create a new industry of AI-native legal service providers.

AI-Powered Real Estate Valuation and Appraisals

The real estate industry has long relied on human appraisers to assess property values, determine loan approvals, and guide investment decisions, an often slow, subjective, and inconsistent process. AI is now making it possible to generate highly accurate property valuations using vast datasets that include market trends, neighborhood analytics, and predictive pricing models.

AI-powered real estate valuation platforms eliminate the need for traditional appraisals by providing instant, algorithm-driven property assessments. These systems use satellite imagery, historical sales data, local economic indicators, and even consumer sentiment analysis to predict property values with a level of accuracy that surpasses human appraisers.

This shift accelerates the home-buying process and enables new financial models. AI-driven fractional ownership platforms allow investors to buy shares in properties based on real-time valuations, creating a new asset class in real estate. In commercial real estate, AI-powered predictive analytics help developers identify the best locations for new properties, reshaping urban planning and investment strategies.

The Next Decade of Market Dominance

The companies that recognize AI's ability to create entirely new value chains will define the next wave of market leaders. Instead of using AI to compete within existing industries, these businesses will use AI to build their own industries.

Just as the rise of the internet gave birth to e-commerce, video streaming, social media, and cloud computing, AI will generate new industries that are not yet fully realized. The key to success in this new landscape is understanding how AI can fundamentally alter who creates value, how it is delivered, and who benefits from it.

For businesses ready to move past incremental improvements and adopt AI as a market-maker, the upcoming decade will shape the future of the industry itself.

The New Competitive Landscape of AI-Native Companies

Companies are integrating AI in different ways. Some optimize existing structures by enhancing workflows, automating processes, and improving productivity. Others build AI into their foundation from the start, creating business models that wouldn't be possible without it. This distinction defines two categories: AI-adapted and AI-native companies.

- AI-adapted companies existed before AI's rise and are now incorporating it into their operations. They modernize legacy systems by layering AI onto traditional processes, gaining effectiveness but facing ongoing disruption from AI-native challengers.

- AI-native companies are designed around AI from day one. They don't just adapt to AI. They are built on it, leveraging its capabilities to create new value, customer experiences, and industry structures. Without legacy constraints, they scale faster and reshape the markets they enter.

Automation vs. Reinvention

The key difference is how these companies apply AI:

- AI-adapted companies use AI to automate and enhance existing processes, such as customer service, predictive inventory management, or machine learning for hiring. While these make operations more efficient, they don't fundamentally change how the business delivers value.

- AI-native companies use AI to create entirely new business models. They harness AI's ability to process massive data, make autonomous decisions, and redefine entire industries rather than simply improving them.

An AI-adapted financial firm may use machine learning to improve fraud detection in credit card transactions. An AI-native financial company, however, may bypass traditional banking altogether, using risk models to underwrite micro-loans in emerging markets where no credit history exists. The AI-adapted firm is making incremental improvements, while the AI-native firm is building a new financial ecosystem.

Why AI-Native Companies Move Faster and Reshape Industries

AI-native companies have structural advantages that allow them to innovate more rapidly and scale more efficiently than AI-adapted businesses. These advantages stem from three core capabilities:

1. **AI-Powered Iteration**: Traditional businesses rely on historical data and periodic strategic planning, then make human-driven adjustments. AI-native companies, however, use AI models that adapt in real time based on incoming data. This allows them to continuously refine their offerings, optimize pricing, and adjust strategies without human intervention.

   Example: An AI-native e-commerce platform can dynamically adjust pricing based on demand, competitor activity, and individual customer behavior, whereas an AI-adapted retailer might only update pricing strategies quarterly based on manual analysis.

2. **Network Effects and Data Advantage**: AI-native companies build ecosystems where data continuously improves AI's capabilities. The more interactions an AI system processes, the smarter it becomes. This creates a competitive flywheel where early adopters of an AI-native platform benefit from increasingly intelligent and efficient services, making it harder for competitors to catch up.

Example: Consider how recommendation engines in streaming services or e-commerce platforms continuously refine their personalization. AI-native companies design their platforms around these learning loops from the beginning, while AI-adapted companies often struggle to retrofit them into existing systems.

3. **Learning-Based Business Models**: AI-native companies apply business models that improve over time rather than remain static. Unlike traditional businesses that offer fixed products or services, AI-native companies deliver solutions that become more valuable the longer they are used.

   Example: In addition to selling software licenses, an AI-native cybersecurity firm offers real-time threat detection that improves with every attack it encounters. A traditional software company might release updates periodically, but an AI-native security platform evolves continuously, making it a more attractive long-term solution.

The gap between AI-adapted and AI-native businesses is growing, and those that fail to evolve risk being left behind. The following comparison illustrates why AI-native companies have a lasting competitive edge and why businesses that only integrate AI into existing structures will struggle to keep up.

Charting the Evolution of Business Models in the AI Era

Feature	Traditional Model (Pre-AI)	AI-Adapted Model (Incremental AI Integration)	AI-Native Model (AI-First Business)
AI's Role	Not central to business operations.	AI enhances existing processes.	AI is the foundation of the business model.
Revenue Model	Fixed pricing, one-time sales, or subscriptions.	AI improves pricing and revenue strategies.	AI enables outcome-based pricing and continuous value delivery.
Customer Interaction	One-size-fits-all products/services.	Personalization for better engagement.	AI predicts customer needs and adapts them in real time.
Competitive Edge	Cost, scale, brand reputation.	AI-driven efficiency improves margins.	AI creates entirely new sources of competitive advantage.
Business Evolution	Slow, dependent on human decision-making.	Faster but limited by existing structures.	Continuous AI-powered iteration with self-learning business processes.
Scalability	Expensive and resource-intensive.	Automation increases productivity.	AI-native businesses scale exponentially through network effects and learning models.
Industry Positioning	Competes within existing markets.	Adapts AI to remain competitive.	Redefines industry boundaries and creates new markets.

By shifting from AI as a tool to AI as a business foundation, companies can create lasting differentiation, new market opportunities, and long-term resilience in an AI-driven world. In the next chapter, we'll explore ways you can position your organization for success in an A-driven future.

Takeaways and Tips

1. **AI is a catalyst for business evolution.** AI should not be limited to cost-cutting or process optimization. Companies that use AI only for automation risk stagnation, while those that employ AI to create new business models and revenue streams will lead their industries.

2. **AI-first businesses redefine competitive advantage.** Companies that embed AI into their core strategy will move faster, scale more efficiently, and reshape market expectations. AI-native businesses do not compete on traditional terms; they create new ones.

3. **Traditional pricing models will become obsolete.** AI enables outcome-based pricing, where customers pay for results rather than static services or products. Companies that transition from fixed pricing to AI-powered performance models will build stronger, more profitable customer relationships.

4. **Ecosystems are replacing traditional value chains.** AI-powered ecosystems create network effects, where value increases as more participants contribute data and interactions. Businesses that build AI-driven platforms will have a long-term advantage over those that operate in isolation.

5. **The gap between AI-adapted and AI-native companies will widen.** AI-adapted businesses may see short-term benefits, but they will struggle to compete with AI-native companies that build business models around AI from the ground up. Organizations that delay this shift risk being overtaken by faster, more adaptable competitors.

Actionable Steps for Competing in an AI-First Economy

- **Communicate with executives.** Through regular leadership updates, reinforce the urgency of moving from efficiency optimization toward strategic innovation.

- **Audit your AI initiatives.** Evaluate if your company is using AI solely for incremental improvements or if it's integrated strategically. Identify and prioritize areas where AI can create new value or disrupt the market.

- **Develop an AI-first strategy.** Conduct workshops to envision how AI could reimagine your company's value. Identify opportunities to move beyond incremental automation into new revenue streams.

- **Form ecosystem partnerships.** Identify potential partners or ecosystems to share intelligence, data, and insights. Begin building or joining AI-powered ecosystems to amplify competitive advantage.

- **Explore outcome-based revenue models.** Test outcome-based pricing strategies, subscription-based AI services, or predictive offerings that continuously evolve and add value.

- **Rethink customer engagement through AI.** AI-native companies predict customer needs before they arise. Explore personalization, anticipatory commerce, and AI-powered services that adapt on the fly.

- **Future-proof your competitive strategy.** Companies that fail to evolve beyond AI-enhanced efficiency will struggle against AI-native competitors. Build a long-term AI roadmap that expands from process improvement to business model reinvention.

CHAPTER 11:
Guiding Your Organization Through AI's Future

Over the next five years, AI will evolve from an optimization tool into a force for innovation. It will integrate seamlessly into operations, redefine competitive advantage, and influence global trends in regulation, workforce development, and sustainability.

Leading voices in AI predict a future of rapid acceleration, with intelligence becoming a new economic driver. Businesses that embrace AI now will optimize their processes and position themselves for a competitive edge in an era of intelligent automation, multimodal AI, and human–AI collaboration. However, with this evolution comes new responsibilities: Organizations must adopt AI ethically, establish governance, and prepare their workforce for the changing landscape.

In this chapter, you'll learn:

- Key AI trends shaping the next decade across industries.

- The shift toward specialized AI solutions and why focused AI tools outperform general models.

- How multimodal AI is transforming business: AI that can see, hear, and understand.

- The effect of AI on the workforce and how automation is reshaping roles, not eliminating them.

- The convergence of AI with technologies like quantum computing and the Internet of Things (IoT) is changing how everyday objects interact, connecting them seamlessly to the internet and unlocking new opportunities for innovation.

In this chapter, we'll explore how understanding these trends can help leaders make informed decisions and harness AI's full potential.

The Look Ahead

The rapid evolution of artificial intelligence is ushering in an era of unprecedented capabilities. Industry leaders agree we are at the cusp of revolutionary change. Sam Altman, CEO of OpenAI, calls this the Intelligence Age, where AI systems will solve problems beyond human capacity. Dario Amodei, CEO of Anthropic, envisions a "compressed 21st century," where AI accelerates progress in fields like biology and neuroscience, achieving a century's worth of breakthroughs in just a decade. Sundar Pichai, CEO of Google and Alphabet, foresees AI seamlessly integrating into daily life, enhancing productivity and addressing global challenges.

From the boardroom to the warehouse floor, AI is reshaping the economy. Altman predicts entirely new categories of better jobs, while Amodei believes AI-powered discoveries, such as curing diseases and extending lifespans, will redefine how we work and live. Pichai takes a more measured view, anticipating a period of rapid advancement followed by steadier progress. Yet all three agree: AI will augment human efforts, enabling workers at all levels to tackle complex tasks more efficiently. AI's influence extends far beyond business. Experts foresee breakthroughs in medicine, healthcare, and environmental science—from compressing decades of research into a few years to addressing food security in vulnerable regions. For business leaders, investing in AI means not only financial returns but also contributions to solving global challenges.

Yet, this future comes with risks. Altman advocates for global collaboration and regulatory guardrails to prevent misuse. Pichai warns of pitfalls like disinformation and unethical applications, while Amodei stresses the need for safe, transparent, and human-aligned AI systems.

For executives navigating this shift, the message is clear: Those who embrace AI as the next great resource will be best positioned to thrive in this evolving landscape.

Key Trends Shaping AI's Future

As we look toward 2030 and beyond, the trajectory suggests a fundamental reshaping of how businesses operate. While overly optimistic predictions should be approached with caution, several key trends are emerging:

1. **Ubiquitous AI Integration**: AI will become embedded in virtually every business process, functioning as a universal capability similar to electricity or internet connectivity. Key examples include:

   - Strategic planning will be enhanced by AI-driven market analysis and forecasting.

   - Operations will be optimized through real-time, predictive adjustments.

- Customer interactions will be personalized at scale via sophisticated AI systems.

2. **Enhanced Human-AI Collaboration**: The organizations that excel will combine human judgment with AI capabilities. This isn't about replacement but amplification:

   - Managers will partner with AI to analyze vast datasets and improve resource allocation.

   - Creative teams will use AI to explore design possibilities and iterate faster.

   - Strategy teams will leverage AI to model complex scenarios and identify opportunities.

3. **Democratized AI Capabilities**: As AI tools become more accessible and user-friendly, the barriers to adoption will fall:

   - Cloud services will enable small and medium-sized businesses to harness enterprise-grade AI capabilities, leveling the playing field so that company size alone no longer determines competitive advantage.

   - Non-technical employees will use natural language interfaces to harness AI tools.

   - Custom solutions will be developed without extensive data science expertise.

The Rise of Focused AI Solutions

As AI continues to evolve, businesses are shifting away from trying to use one massive system to handle everything. Instead, they are focusing on smaller, more specialized AI solutions designed to solve specific problems exceptionally well. This is like hiring a highly skilled expert for a particular task rather than a jack-of-all-trades who might not excel at anything specific.

Because every industry faces unique challenges, specialized AI tools are built with these differences in mind and created to address particular needs within industries like healthcare, finance, or manufacturing.

For leaders like you, specialized AI offers a clear path to better outcomes without unnecessary complexity. Here's why this shift is important:

1. **More Accurate Results:** Specialized AI tools are often better at solving specific problems than general-purpose systems. For example, a healthcare AI trained specifically on heart conditions will likely outperform a broader tool that tries to handle all types of medical diagnoses.

2. **Faster and Cheaper to Implement:** These tools are often quicker to set up because they're built for a specific use case. This means your team spends less time (and money) adapting the system to your business.

3. **Better Fit for Your Goals:** Specialized AI integrates closely with your company's key priorities, whether it's improving customer satisfaction, reducing costs, or increasing performance. You're not paying for extra features you don't need.

4. **Easier to Grow With Your Business:** As your business expands or faces new challenges, you can add focused AI solutions that address specific needs rather than overhauling a giant, general system.

Specialized AI solutions work like precision instruments, helping your team focus on what really matters. They're faster to implement, easier to use, and more effective at solving the specific problems that affect your bottom line. By thinking of AI as a collection of smart, focused tools rather than a one-size-fits-all system, you'll be better positioned to unlock its full potential for your company.

Entering the Truly Native Multimodal Era

Artificial intelligence is evolving to become more like how humans interact with the world. Until now, many of your interactions with AI might have been through text, like using chatbots for customer service, virtual assistants like Siri or Alexa, or automated responses in your email system. The future of AI goes beyond understanding and generating text. We're entering the multimodal era, where AI can also process and understand images, videos, sounds, and data from various sensors. Additionally, advancements in AI are introducing world models: comprehensive internal representations that help AI systems understand and interact with the world in a more integrated and intuitive way.

A More Human-Like Understanding

Imagine AI that moves beyond reading and writing to see, hear, and interpret the world around it. This new generation of AI can handle multiple types of information at once, making it much more versatile and powerful. These systems and world models can build a cohesive understanding of their environment, enabling smarter decision-making and action. Here are a few examples:

- **Smart Factory Monitoring:** AI systems can "watch"

your factory floor through cameras and sensors, spotting issues or unusual patterns in real time. With world models, AI can understand the context of these anomalies, predicting potential breakdowns before they happen and suggesting optimal maintenance schedules.

- **Enhanced Virtual Assistants:** Picture virtual assistants that respond to your spoken commands, display visual information like charts and dashboards, and remember prior requests you've made. With world models, these assistants can better understand the context of your requests, providing more relevant and insightful responses.

- **Optimized Retail Operations:** In retail, AI can analyze video feeds from store shelves to confirm products are placed in the best way, helping to increase sales and improve customer satisfaction. World models allow the AI system to integrate this visual data with sales trends and inventory levels, providing a holistic view of store performance.

Implications for Your Business

Multimodal AI, empowered by world models, brings your operations a new level of interaction and precision. Here's how it can make a difference:

1. **Natural Interactions:** In meetings and presentations, your team can ask questions out loud during meetings and instantly see relevant data visualizations on the screen. World models enable AI to understand the context of the discussion, providing more accurate and meaningful data insights.

2. **Enhanced Field Operations**: Field technicians can use smart glasses or other wearable devices to guide them through complex repairs with visual instructions and real-time data. This is all informed by a comprehensive world model that anticipates potential challenges and has learned from past repair outcomes.

3. **Improved Customer Engagement**: Customers can interact with AI more naturally, using voice and visuals, making their experience smoother and more satisfying. World models help AI understand customer needs in context, leading to more personalized service.

4. **Advanced Employee Training**: Videos and interactive simulations can provide more effective training for your employees, helping them learn faster and retain information better. With world models, training scenarios are realistic, relevant to actual job conditions, and personalized to the individual's own learning pace.

5. **Superior Quality Assurance:** AI can continuously monitor production lines with visual and sensor data, boosting quality standards and quickly addressing any safety issues. World models enhance AI's ability to predict and prevent defects by understanding the broader production context.

The New Workforce Dynamic

As artificial intelligence continues to advance, it is evolving into intelligent systems that can perceive and interact with the physical environment.

Your New Coworker: AI Agents

When AI-driven insights become easily accessible across the organization, management responsibilities naturally become more distributed among teams. Organizational structures flatten out, empowering employees at every level to make informed, strategic decisions. Leadership shifts from a traditional top-down approach to a more collaborative. This change means people won't just need to be good at interpreting data; they'll also need to lean into critical thinking, creativity, emotional intelligence, and strong communication skills. As technology continues evolving quickly, continuous learning and adaptability will become essential for everyone.

Robotics: Integrating Machines and Humans

Robots today are equipped with advanced sensors, cameras, and microphones that enable them to see, speak, and communicate, turning them into active participants on the factory floor, in warehouses, and throughout various business settings.

These next-generation robots can monitor their surroundings with a level of situational awareness that was previously unattainable. For example, a robot can now detect if a tool or critical component is misplaced, instantly alerting human operators to potential issues before they disrupt the workflow. This ability to visually confirm the presence and proper placement of tools not only improves efficiency but also enhances safety by preventing accidents and reducing downtime.

Along with their visual perception, the integration of natural language processing enables these robots to comprehend and carry out verbal instructions, effectively bridging the divide between human intent and machine action. They can participate in two-way communication with human colleagues, providing essential information such as inventory updates, maintenance needs, or safety alerts. This seamless communication cultivates a collaborative environment where people and robots work together.

A Collaborative Future

Organizations that successfully blend human ingenuity with machine optimization will stand out in the competitive landscape. By embracing this collaborative approach, your workforce will remain adaptable, motivated, and equipped to tackle the challenges of the future.

Fusing AI with Emerging Technologies

Artificial intelligence isn't developing in isolation. It will increasingly integrate with other groundbreaking technologies, such as quantum computing, advanced sensors, 5G/6G networks, and immersive AR/VR platforms. This convergence will unlock many breakthroughs.

Here's how these technologies will work together to create powerful new solutions:

- **Quantum Computing:** When combined with AI, quantum computing has the potential to process information at speeds unimaginable to today's classical computers. For example, AI algorithms running on classical computers, tasked with complex problems such as factoring large encryption keys, could require thousands or even millions of years. However, the same AI algorithms, empowered by quantum computing and Shor's algorithm, could solve these tasks within hours. Practical applications include financial institutions using quantum-powered AI to conduct real-time risk assessments and optimize investment strategies, while pharmaceutical companies can speed up drug discovery by simulating molecular interactions with unmatched accuracy.

- **Advanced Sensors:** Modern sensors can collect vast amounts of data from the physical world, feeding AI systems with real-time information. In manufacturing, advanced sensors integrated with AI can monitor equipment health, predict maintenance needs, and prevent costly downtimes. Similarly, in agriculture, sensor-driven AI can analyze soil conditions, weather patterns, and crop health to strengthen farming practices and increase yields.

- **5G/6G Networks:** The next generation of wireless networks will offer high-speed, low-latency connectivity essential for real-time AI applications. In remote field operations, such as oil rigs or construction sites, 5G-enabled AI systems can provide immediate insights and aid decision-making processes. Improved connectivity also allows seamless integration of AI with Internet of Things (IoT) devices, creating smarter and more responsive environments.

- **AR/VR Platforms:** Augmented Reality (AR) and Virtual Reality (VR) technologies can enhance AI's capabilities by providing immersive data visualization and interactive experiences. In retail, AR-enabled AI can offer customers virtual try-ons and personalized shopping experiences. In training and education, AI-powered VR simulations can create realistic scenarios for employee development, improving learning outcomes and skill retention.

One thing is clear: the AI landscape is constantly evolving. Breakthroughs don't just shift the conversation—they rewrite the rules. What's cutting-edge today could be obsolete tomorrow. The leaders who recognize this—and act with clarity, speed, and curiosity—won't just keep up; they'll set the pace.

Takeaways and Tips

1. **AI is shifting from hype to real-world transformation.** It is no longer an emerging trend but a key driver of business strategy, competitive advantage, and industry innovation. Organizations that proactively integrate AI will lead in efficiency, decision-making, and market positioning.

2. **Specialized AI solutions will outperform general-purpose systems.** Companies investing in AI solutions tailored to specific business problems will achieve better results, faster implementation, and higher ROI compared to broad, generalized AI tools.

3. **Multimodal AI will redefine how businesses interact with technology.** AI has already evolved beyond text-based models to incorporate images, speech, and real-world data, creating more human-like interactions. Companies should explore how multimodal AI can enhance customer engagement, operations, and analytics.

4. **A workforce revolution is taking place.** AI will take over repetitive and routine tasks, allowing employees to focus on creativity, strategy, and problem-solving. Businesses should invest in reskilling programs to prepare their workforce for AI-augmented roles, as everyone will become a manager of agents.

5. **AI will merge with other breakthrough technologies.** The convergence of AI with quantum computing, IoT, 5G, and AR/VR will unlock entirely new capabilities. Forward-thinking businesses should monitor these developments and explore how they can be utilized for long-term growth.

Actionable Steps to Prepare for the AI-Powered Future

- **Future-proof your organization**. Develop and regularly update a roadmap outlining how your organization will adapt to AI-driven changes over the next 5 to 10 years. Cultivate an innovation-focused culture, ensuring your teams are prepared to embrace ongoing technological shifts and experimentation.

- **Foster technological synergies**. Invest in strategic partnerships and R&D projects that integrate AI with quantum computing, IoT sensors, 5G/6G networks, and AR/VR technologies. Continuously monitor technological advancements through industry events, peer benchmarking, and thought leadership networks.

- **Invest in human-AI collaboration**. Develop training programs that enhance collaboration between people and AI, prioritizing adaptability, creativity, critical thinking, and emotional intelligence. Clearly communicate to your workforce the complementary nature of AI and proactively address potential resistance.

- **Monitor the evolution of AI and emerging technologies**. Stay informed about how AI is converging with other advancements and evaluate potential strategic applications for your company.

CONCLUSION:
It's Just the Beginning

As this book draws to a close, calling it a "Conclusion" feels inadequate. Over the past year, as I wrote, edited, and rewrote this book, AI continued to advance at a staggering pace, reshaping industries and expanding the limits of what's possible. Deciding when to stop was a challenge because there was always something new to explore. But rather than viewing this as the final word, consider it an open invitation to engage, experiment, and influence what comes next.

The AI landscape is anything but static. Breakthroughs emerge constantly, shifting the conversation and redefining best practices. Strategies that seem cutting-edge today may soon require rethinking, and leaders who embrace this reality will be the ones who stay ahead. Curiosity and adaptability aren't optional; they are essential. Those willing to evolve with AI will not only navigate the changes but shape them.

This is the moment to put ideas into motion. Test them. Challenge them. Let discussions with peers sharpen your perspective. AI is not a road with a fixed destination but an unfolding journey, one that rewards those who stay engaged and open to discovery. The future belongs to those who don't just follow progress but help define it.

Next Steps

The journey through these pages should equip you with a practical understanding of where AI can fit into your business and how to chart a path forward. The next step is to turn insight into action. To integrate AI into your business in a meaningful and sustainable way, consider adopting a structured, iterative approach:

1. **Identify key opportunities**. Pinpoint where AI can deliver the most immediate value. This may include optimizing supply chains, improving customer support through chatbots, enhancing forecasting accuracy, or accelerating research and development. Choose low-risk but high-impact use cases to validate AI's potential within your specific business context.

2. **Establish a cross-functional AI team**. Build a dedicated team that blends technical acumen with domain expertise. Include data scientists, software engineers, and business analysts, as well as representatives from operations, marketing, compliance, and customer service. This will ensure that your AI initiatives are technologically feasible and aligned with organizational objectives.

3. **Develop a clear road map and governance model**. Outline a phased plan for AI adoption. Begin with small pilot projects, assess results, refine methods, and then scale up. Adopt governance frameworks to protect data quality, privacy, security, and compliance. This establishes a stable foundation upon which your AI solutions can grow responsibly.

4. **Invest in tools, training, and culture**. Provide your team with access to cutting-edge platforms and technologies and commit to ongoing skill development. Cultivate a culture that encourages experimentation, learning from failure, and quick adaptation. Empower employees at all levels to grasp and utilize AI-driven insights in their daily tasks and decision-making.

5. **Measure outcomes and iterate**. Track the performance of your AI initiatives through carefully chosen metrics: efficiency improvements, revenue growth, customer satisfaction, and risk reduction. Use these insights to fine-tune your models, update workflows, and identify new areas where AI can add value. Continuous iteration is the essence of successful AI integration.

6. **Stay informed and adaptable**. The AI landscape evolves quickly. Keep pace with advancements in algorithms, infrastructure, ethics, and regulation. Regularly reassess your portfolio of AI projects to guarantee they remain relevant to strategic business goals.

By acting on these recommendations, you can move beyond theory and position your organization to seize the opportunities AI offers. The goal is to elevate your organization's capabilities by improving decision-making, delighting customers, and ultimately achieving a sustainable competitive advantage. AI is poised to become a critical lever in shaping your business's future; now is the time to make it happen.

Let's keep the conversation going. The future of AI is unwritten, and together, we have the power to shape it.

GLOSSARY

Adaptive Products

Products that learn from user interactions and adjust features or functionality over time.

AI Champion

An internal advocate responsible for promoting AI adoption, fostering collaboration, and making sure AI strategies and organizational goals match.

AI Ethicist

A professional dedicated to ensuring AI systems align with ethical, societal, and organizational standards, promoting fairness, transparency, and accountability.

AI Governance Council

A multidisciplinary team of stakeholders responsible for overseeing the ethical, strategic, and operational implementation of AI initiatives.

AI Governance

A framework of policies, practices, and procedures regulating the ethical, secure, and efficient deployment of AI technologies.

AI Philosophy

A set of guiding principles ensuring that AI use fits with a company's values, strategic objectives, and societal responsibilities.

AI Pilot Programs

Small-scale, controlled AI implementations designed to test feasibility, assess effects, and evaluate scalability before full deployment.

AI Unicorn

A rare leader proficient in technical AI knowledge, strategic planning, and interpersonal skills to drive effective AI initiatives.

AI Wrapper Solutions

Add-on tools that embed AI capabilities into existing processes or systems without requiring structural overhauls.

Algorithmic Fairness

The practice of designing AI systems that do not perpetuate biases or inequalities so decision-making remains equitable and impartial.

Artificial General Intelligence (AGI)

Advanced AI capable of understanding, learning, and performing intellectual tasks across multiple domains, comparable to human cognition.

Artificial Intelligence (AI)

The science of creating systems and machines capable of learning, reasoning, and adapting to perform tasks traditionally requiring human intelligence.

Automation

Technology that executes repetitive tasks based on pre-programmed rules without learning or adapting to new conditions.

Bias Audit

A systematic review of AI systems to identify and address potential biases in their algorithms or data.

Bias Detection Algorithms

Algorithms that identify and mitigate biases in data or models so AI outputs are fair and accurate.

BYOAI (Bring Your Own AI)

A trend where employees bring personal AI tools into workplaces, affecting organizational security, governance, and strategy.

Chain of Thought Model

An AI reasoning framework that breaks down complex problems into a series of logical steps, improving accuracy and clarity in decision-making.

Change Management

The structured process of guiding individuals and teams through organizational changes to minimize resistance and maximize success.

Commoditization

The process by which AI tools become widely accessible and standardized, requiring businesses to innovate for differentiation.

Copyright Protection

Legal frameworks safeguarding the ownership and usage rights of creative works, including AI-generated content.

Core Processes

The essential workflows and operations that form the foundation of an organization's success and competitiveness.

Custom GPTs

Personalized language models tailored for specific tasks, industries, or organizational needs, enabling more relevant and accurate outputs.

Data Governance

A set of practices and policies ensuring that data is managed ethically, securely, and effectively across its lifecycle.

Data Governance Framework

A structured approach defining roles, responsibilities, and guidelines for managing data within an organization.

Data Literacy

The ability to interpret, analyze, and communicate data insights effectively to make informed decisions.

Data-Driven Decision-Making

The use of data insights and analytics to guide decisions.

Deepfake

AI-generated content that manipulates audio, images, or video to create realistic likenesses but fabricated media.

Ecosystem Thinking

A holistic strategy that emphasizes interconnected relationships and collaboration across stakeholders, contrasting with linear value chain models.

Entrepreneurial Operating System (EOS)

A framework of tools and practices designed to help businesses align goals, streamline operations, and improve organizational health.

Ethical AI

AI designed and implemented with a focus on fairness, transparency, and societal impact.

Exposure Key

A framework for mapping organizational tasks and work-flows to potential AI capabilities, identifying automation and enhancement opportunities.

Fair Use Policy

Guidelines permitting limited use of copyrighted material without permission, balancing innovation with respect for IP rights.

Fractional AI Leadership

Part-time or advisory AI leadership roles that provide strategic guidance without requiring full-time employment.

GenAI Native Solutions

AI tools explicitly designed to address specific business challenges, leveraging generative AI technologies.

Generative AI

AI systems capable of creating new content, such as text, images, or music, based on existing data and patterns.

Generative Design

AI-driven tools that explore multiple design possibilities based on input criteria, optimizing for performance and creativity.

Human-Centered AI

AI designed with a focus on ethical considerations, user trust, and empowering human decision-making rather than replacing it.

Hybrid Model

An organizational approach where AI initiatives are managed centrally but also embedded within individual departments.

Indemnification

A legal agreement that protects one party from financial losses or damages caused by another party's actions.

Intellectual Property (IP)

Creations of the mind, such as inventions, designs, or works of art, legally protected to grant ownership rights.

Intelligent Automation

The combination of AI and automation to handle complex tasks requiring cognitive capabilities with minimal human intervention.

JobsGPT

An AI tool designed to analyze job descriptions and identify opportunities to integrate AI.

Knowledge Graph

A data structure representing relationships between entities, enabling AI to connect and interpret information contextually.

Machine Learning (ML)

A subset of AI that enables systems to learn from data, improve over time, and make predictions without explicit programming.

Micro-Segments

Highly specific customer groups identified through AI-driven analysis of behaviors and preferences.

Moonshots

Ambitious, high-impact AI initiatives requiring significant investment and long-term commitment to achieve transformational results.

Multimodal AI

AI systems capable of processing and interpreting multiple types of input, such as text, images, and audio, simultaneously.

Natural Language Generation (NLG)

AI technology that converts structured data into coherent, human-like text, useful for reporting or content automation.

Natural Language Processing (NLP)

The ability of AI to understand, interpret, and respond to human language through text or speech.

Neural Networks

Algorithms modeled after the human brain, enabling pattern recognition, learning, and decision-making in AI systems.

Open vs. Closed Models

Comparison of AI systems based on accessibility. Open models allow customization and broad access. Closed models prioritize security and control over proprietary technology.

Predictive Analytics

The use of historical data and algorithms to forecast future outcomes, aiding proactive decision-making.

Process Mapping

A technique for visualizing workflows to identify inefficiencies, redundancies, and optimization opportunities.

Psychographic Analysis

Analysis of people's attitudes, values, and behaviors to better understand target audiences for marketing or product development.

Quantum-Safe Cryptography

Encryption methods designed to withstand potential threats posed by quantum computing's advanced processing power.

Reinforcement Learning

An AI training technique where systems learn optimal actions through feedback and rewards for successful outcomes.

Retrieval-Augmented Generation (RAG)

A method combining domain-specific data with language models to improve the relevance and accuracy of AI outputs.

Robotic Process Automation (RPA)

Automation of repetitive, rule-based tasks using software robots, freeing up human workers for higher-value activities.

Sentiment Analysis

AI tools used to detect and interpret emotions or opinions expressed in text, enabling better customer understanding.

Stakeholder Panels

Groups representing diverse perspectives, providing ongoing feedback to ensure AI initiatives meet ethical and practical goals.

Web Scraping

Automated data collection from websites, often used for training AI models or gathering competitive intelligence.

Weights and Biases

Parameters in AI models that determine the importance of input features, shaping the system's predictions and outputs.

World Models

Internal representations created by AI systems to simulate and understand their environment, enabling smarter decision-making.

APPENDIX:
Essential Resources for Executives in the AI Era

In the dynamic world of artificial intelligence, staying informed is essential for executives aiming to make sound strategic decisions. Leaders can stay ahead of the curve by leveraging curated content, thought leadership, and diverse learning formats. Here's a guide to resources for keeping pace with AI advancements.

1. **Subscribe to specialized news sources and newsletters**. With AI evolving rapidly, staying updated can feel overwhelming. Here are a few top recommendations:

   - **Demystifai Digest**: Business-oriented news on AI, curated for executives.

   - **Marketing AI Institute**: Hosts MAICON, "the one conference dedicated to marketing AI."

   - **The Information**: A technically focused resource for staying informed on cutting-edge developments.

   - **MIT Technology Review**: Offers in-depth analysis of AI advancements, including its annual list of breakthrough technologies.

   - **The Verge AI**: Comprehensive coverage of machine learning, robotics, and ethical implications of AI.

- **Ethan Mollick's One Useful Thing Newsletter**: Translates complex AI topics into actionable ideas, providing insights on integrating AI into organizational strategies.

2. **Follow AI leaders on social media.** Business leaders looking to stay informed about AI developments should follow the following influential figures in the field.

Tech Industry Leaders:

- **Sam Altman**, CEO of OpenAI

- **Sundar Pichai**, CEO of Google

- **Mark Zuckerberg**, CEO of Meta

- **Satya Nadella**, CEO of Microsoft

- **Jensen Huang**, CEO of NVIDIA

AI Researchers and Pioneers:

- **Geoffrey Hinton**, known as the "Godfather of Deep Learning"

- **Yann LeCun**, Chief AI Scientist at Facebook

- **Andrew Ng**, co-founder of Coursera and former Chief Scientist at Baidu

- **Fei-Fei Li**, co-director of Stanford's Human-Centered AI Institute

- **Demis Hassabis**, co-founder and CEO of DeepMind

AI Ethics and Policy Advocates:

- **Timnit Gebru**, leading researcher in AI ethics

- **Stuart Russell**, professor at UC Berkeley

- **Amandeep Singh Gill**, United Nations Envoy on Technology

Industry Innovators:

- **Dario Amodei**, CEO of Anthropic, working on advanced AI systems

- **Aravind Srinivas**, CEO of Perplexity, innovating in AI-powered search and information retrieval

- **Victor Riparbelli**, CEO of Synthesia, pioneering AI-generated video content

3. **Listen to informative podcasts**. Podcasts offer a flexible way to absorb expert knowledge. Alongside Paul Roetzer's Artificial Intelligence Marketing Institute Podcast, consider these options:

- **AI Weekly**: A broad overview of weekly AI advancements.

- **The VC20**: Insights into AI from a venture capital perspective.

- **Lex Fridman Podcast**: In-depth discussions on AI ethics, robotics, and innovation.

- **a16z**: An exploration into AI's transformative value across industries.

4. **Participate in webinars and virtual events**. Interactive learning opportunities like webinars and virtual conferences provide direct access to AI experts. These events often feature live presentations, panel discussions, and Q&A sessions, making them ideal for exploring specific topics in depth.

5. **Take online learning courses**. Structured online courses help executives deepen their AI expertise. Many programs are tailored for business leaders and cover foundational knowledge and advanced applications in organizational strategy.

By using the various resources listed here, executives can cultivate a well-rounded understanding of AI. Staying informed empowers leaders to make strategic decisions and unlock AI's full potential. For a curated list of resources, visit:

https://www.demystifai.com/aimadesimple/execlibrary

A Daily Integration Approach for Executives

The most effective way for non-technical executives to stay updated on AI advancements is to embed AI tools and practices into everyday professional and personal activities.

In the workplace, AI-powered solutions such as intelligent project management systems and data-driven decision-making platforms can streamline operations and enhance strategic insights. Regular use of these tools boosts productivity and familiarizes executives with the latest AI capabilities, enabling them to make informed decisions.

Beyond the professional sphere, integrating AI into your personal life reinforces understanding and proficiency with the technology. Using digital personal assistants for managing schedules, adopting smart home devices, and employing AI-enhanced tools for hobbies can demystify AI's practical applications. This hands-on experience fosters a comfortable and intuitive relationship with AI, making integrating it into business practices more seamless. Additionally, staying informed through AI-focused news, participating in relevant communities, and continuously experimenting with new AI applications help you stay informed about emerging trends.

The landscape of artificial intelligence is vast and continually evolving. By staying curious and inviting AI into every aspect of your organization, you position your company to lead the way in innovation and strategic growth. Embrace the journey and use AI's strengths to drive your organization toward a prosperous and dynamic future.

ACKNOWLEDGMENTS

First and foremost, I want to express my heartfelt gratitude to Shadley Grei, whose patience and clarity helped transform the chaos in my brain into coherence on the page.

To my husband, Nick Becraft, whose unwavering support made this work possible, and to my children, who so graciously shared their mommy with her "robot friends" during countless hours of writing and research.

To my colleagues at the Jones School at Rice University, especially Amit Pazgal and Barbara Bennett, whose expertise and encouragement proved invaluable. To my friends, particularly Alison Lami, and to my dad, who willingly served as guinea pigs for early ideas and drafts. Your candid feedback shaped this book in countless ways.

And finally, a special nod to my tireless AI collaborator, the relentless digital companion who continually challenged and pushed my thinking, helping me unlock ideas trapped in my brain and proving that even artificial minds can teach us profound lessons about leadership in this new technological era.

ABOUT THE AUTHOR

Kathleen Perley is a recognized AI thought leader, entrepreneur, and educator, bridging the gap between AI and executive decision-making. As the founder of DemystifAI and an AI advisor at Rice Business, she consults with Fortune 500 companies, boards, and leadership teams on AI strategy, implementation, and governance. She specializes in turning AI complexity into real-world business impact.

www.ingramcontent.com/pod-product-compliance
Lightning Source LLC
Chambersburg PA
CBHW071553210326
41597CB00019B/3224